Combined Science

Required Practicals

Exam Practice Workbook

Primrose Kitten

GCSE

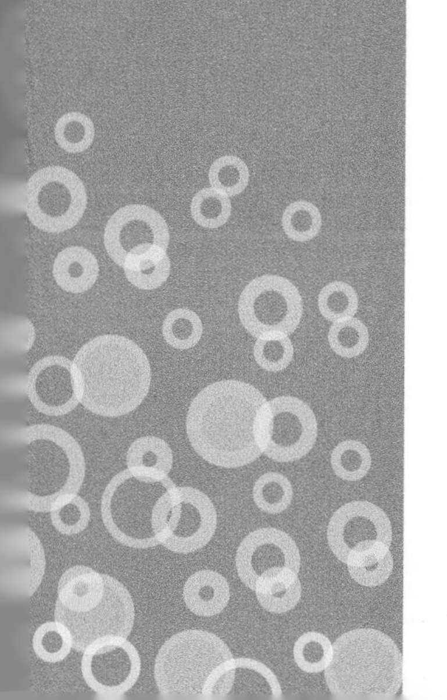

Great Clarendon Street, Oxford, OX2 6DP, United Kingdom

Oxford University Press is a department of the University of Oxford. It furthers the University's objective of excellence in research, scholarship, and education by publishing worldwide. Oxford is a registered trade mark of Oxford University Press in the UK and in certain other countries

First published in 2019

British Library Cataloguing in Publication Data
Data available

978 0 19 844492 3

10 9 8 7 6 5 4 3

Paper used in the production of this book is a natural, recyclable product made from wood grown in sustainable forests.
The manufacturing process conforms to the environmental regulations of the country of origin.

Printed in Great Britain by Ashford Colour Press Ltd. Gosport

Acknowledgements

COVER: ERICH SCHREMPP/SCIENCE PHOTO LIBRARY

Artwork by Aptara Inc.

Contents

Introduction

As part of your AQA GCSE Combined Science course, you will carry out 21 Required Practicals. You can be asked about any aspect of any of these during the exams; this can include planning an investigation, making predictions, taking readings from equipment, analysing results, identifying patterns, drawing graphs, or suggesting improvements to the method. You can also be asked about practicals that are similar but that you may not have done before. You need to be able to recognise and apply the key practical skills that you have learnt to different experiments.

These practical questions account for at least 15% of the total marks. This Exam Practice Workbook allows you to practise answering questions on the 21 Required Practicals and become familiar with the types of questions you may find in the exams. There are lots of hints and tips about what to look out for when answering practical questions. Answers to all questions are available at oxfordsecondary.co.uk/required-practicals-answers.

Practical method – Full details of all 21 Required Practicals, including equipment, method, and safety information, will remind you of the practical work you have carried out and the important skills you have gained during the course

Exam tips – Hints on how you can approach the practical exam questions, improve your answers, and secure marks

5 Photosynthesis

Investigate the effect of light intensity on the rate of photosynthesis.

Method

1 Cut a 10 cm piece of pondweed.
2 Place the piece of pondweed into a beaker of water, covered with an inverted filter funnel. Make sure the cut end of the pondweed is at the top.
3 Fill a measuring cylinder with water and carefully invert it over the top of the filter funnel.
4 Position a lamp exactly 100 cm from the pondweed, switch the lamp on, and leave it for two minutes to allow the pondweed to acclimatise.
5 Start a stopclock and record the number of bubbles produced in three minutes.
6 After 3 minutes, record the volume of gas that has been collected in the measuring cylinder.
7 Refill the measuring cylinder with water and repeat steps 4–6 for distances of 80 cm, 60 cm, 40 cm, and 20 cm.

Equipment

- pondweed
- scissors
- lamp
- large beaker
- filter funnel
- 10 cm^3 measuring cylinder
- metre ruler
- stop clock

oxygen
water
oxygen bubble
pondweed

Safety

- Wash hands after contact with pondweed and pond water.
- Lamp may get hot.
- Keep electrical equipment dry and do not handle if hands are wet.
- Dispose of the pondweed responsibly.

Remember

The skill being tested in this practical is your ability to measure and calculate rates of photosynthesis. You should be able to describe how to measure the rate of a reaction or process by collecting a gas that is produced.
Remember that you will probably have plotted a graph of the volume of gas collected (or number of bubbles) against the distance between the lamp and the pondweed (or light intensity). It is important to know what this graph looks like and to be able to explain its shape.

Exam Tip

- A living plant is needed for this experiment to work properly. Plants only undergo photosynthesis in the right conditions, and a number of factors need to be controlled. The following questions will change some of those factors, some in an obvious way and some in a not so obvious way.
- You need to know the word equation for photosynthesis. You should also be able to recognise the chemical formulae for all the substances in the equation.

1 Identify the dependent variable in this experiment. [1 mark]

2 Draw one line from each compound to its chemical formula. [3 marks]

carbon dioxide	$C_6H_{12}O_6$
glucose	H_2O
oxygen	CO_2
water	O_2

Exam Tip

In science exams, you should make your points in as few words as you can. Your answer for this question should be no more than five words long. Do not answer this in a full sentence and do not repeat the question.

3 Complete the word equation for photosynthesis. [4 marks]

_____ + _____ $\xrightarrow{\text{light}}$ _____ + _____

4 Identify the gas in the bubbles produced at the cut end of the pondweed. [1 mark]

5 Two students carried out this practical with different light sources.
- Student A used a desk lamp.
- Student B used the light on their mobile phone.

Suggest how you would expect the students' results to differ and explain why. [3 marks]

30 31

Remember – Each practical has a reminder of the key skills being tested in the practical, whatever the context

Exam-style questions – Lots of practical exam-style questions about each Required Practical help you to become confident in answering practical questions

Table of Required Practicals

The Required Practicals in this workbook are numbered and ordered according to the Combined Science: Trilogy specification. The practicals are the same for Combined Science: Synergy, as shown in the table below. Complete the checklist as you go through the workbook to show how confident you are with each practical.

Required Practical	Trilogy	Synergy	Checklist ☹	Checklist 😐	Checklist ☺
Microscopy	1	3			
Osmosis	2	4			
Food tests	3	7			
Enzymes	4	20			
Photosynthesis	5	10			
Reaction time	6	8			
Field investigations	7	12			
Making salts	8	17			
Electrolysis	9	21			
Temperature changes	10	18			
Rates of reaction	11	19			
Chromatography	12	9			
Water purification	13	11			
Specific heat capacity	14	2			
Resistance	15	16			
I–V characteristics	16	15			
Density	17	1			
Force and extension	18	13			
Acceleration	19	14			
Waves	20	5			
Radiation and absorption	21	6			

1 Microscopy

Use a light microscope to observe, draw, and label biological specimens.

Method

Before you can look at the cells on the slide, you will need to set up your microscope.

Most microscopes have a built-in light source, but if the one you are using does not then you need to arrange the mirror found underneath the stage so that light is directed through the lens system.

1. Move the stage to its lowest position.
2. Place a prepared slide on the centre of the stage and fix it in place using the clips.
3. Select the objective lens with the lowest magnification and raise the stage to its highest position.
4. Look through the eyepiece and slowly move the stage down by turning the coarse focus adjustment until the cells on the slide come into view.
5. Turn the fine focus adjustment to sharpen the focus so the cells can be clearly seen.
6. If you wish to view the object at greater magnification to see more detail, switch to a higher magnification objective lens and use the fine focus adjustment to sharpen the focus.

Safety

- Take care when handling glass slides as they are very fragile.
- Take care not to break the slide by moving the stage too close to the objective lens.

Equipment

- light microscope with low and high power objective lenses
- a range of prepared animal cells including:
  - cheek cells
  - red blood cells
- a range of prepared plant cells including:
  - onion epidermal cells
  - leaf palisade cells

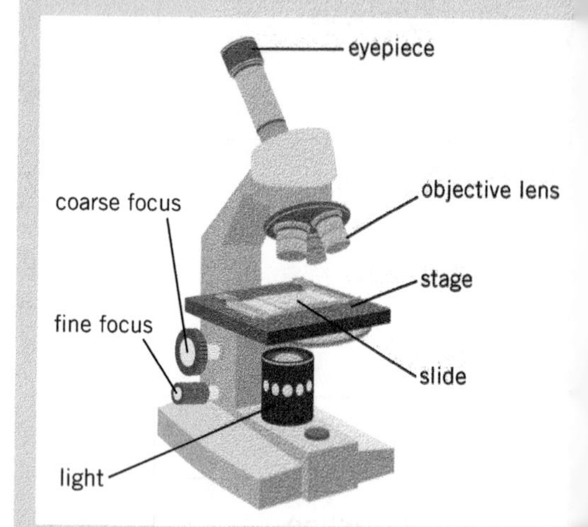

Remember

The skill being tested in this practical is whether you can use a light microscope to observe plant and animal cells. You need to be able to describe how to set up the microscope, focus on a slide containing the specimen, and then make a labelled scientific drawing of what you see. Don't forget to include the magnification in scientific drawings.

To calculate the total magnification of the microscope you used to see your cells:

total magnification = eyepiece lens magnification × objective lens magnification

Exam Tip

In the exam, you may be asked about a specimen you haven't come across before. It's important to remember that it's still just a plant or animal cell. Apply your knowledge of the microscopy practical to the exam question, whatever the specimen.

1 Label the diagram of an animal cell. [5 marks]

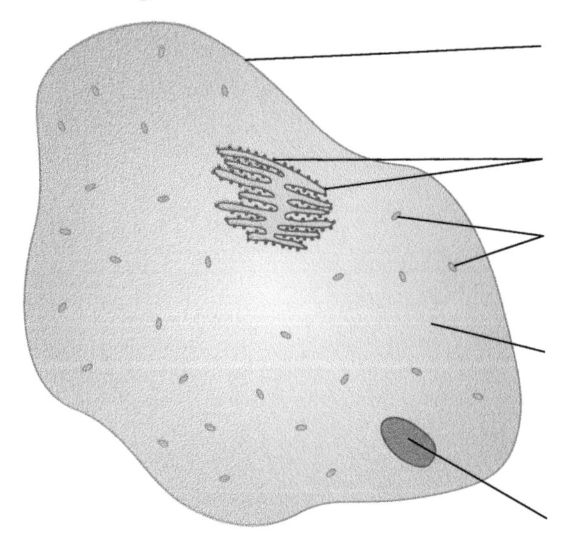

2 Complete the diagram of a plant cell by giving the functions of the labelled organelles. [5 marks]

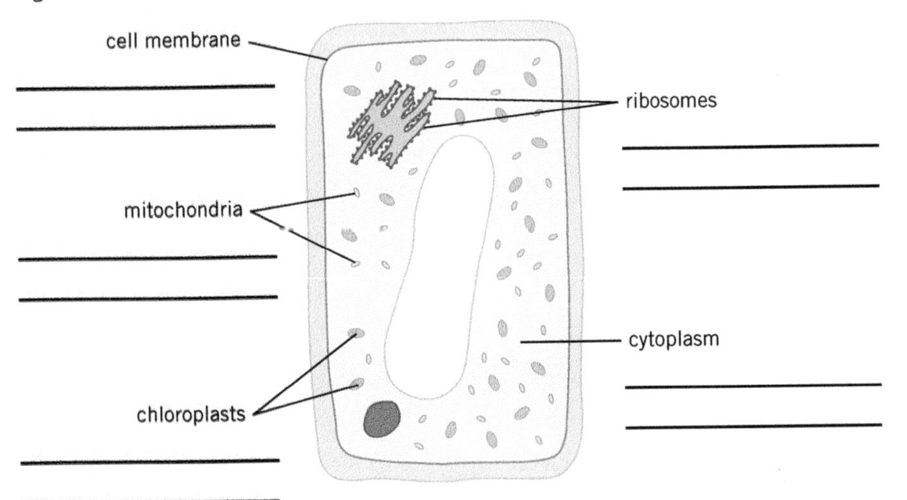

3 A student examines an onion cell under a microscope. Suggest why this plant cell is not green. [1 mark]

4 Draw **one** line from each part of the microscope to its function. [4 marks]

Part	Function
Slide	Part that can move around, so you can view different sections of the sample
Eyepiece	Smooth curved piece of glass closest to the sample being viewed
Objective lens	Piece of glass where the sample is immobilised and stained
Stage	Part of the microscope that you look through

5 Suggest why it is not possible to see the internal structures of a bacterial cell using a light microscope. [1 mark]

____

____

____

____

6 Which objective lens should you use when you first focus a microscope on a slide? [1 mark]

Tick **one** box

×4 ☐

×40 ☐

×100 ☐

×400 ☐

Exam Tip

For this question you need to select ONE answer.

You will not get any marks if you tick more than one box, even if one of the boxes you tick is correct.

7 Suggest why it is important to state the magnification on any drawings you make from a microscope. [1 mark]

____

____

____

____

8 A student drew three images of a plant cell at three different magnifications but failed to label the drawing with the magnification. The microscope has three objective lenses ×40, ×100, and ×400.

A B C

Draw **one** line from each image letter to the correct objective lens. [3 marks]

Image A	×40
Image B	×100
Image C	×400

9 A student has prepared a sample on a slide and wants to view it using the ×100 objective lens.

Describe the steps the student should take to focus the microscope. [4 marks]

10 The two different classes of microscopes are light microscopes and electron microscopes.

Compare these two different classes of microscopes. [6 marks]

Hint

When 'compare' is the command word in a question, you need to give similarities and differences.

11 Put these objects in order of size, **smallest** first. [2 marks]

A Animal cell (10 µm)

B Ant (1 mm)

C DNA (10 nm)

D Bacterial cell (1 µm)

E Virus (100 nm)

	E			B

Exam Tip

When converting between units, the answer is going to be a 'sensible' size of number. If the answer you get is massive or tiny, you may have multiplied numbers when you needed to divide, or divided numbers when you needed to multiply them.

12 A sample was measured to be 125 000 µm.

Calculate this size in cm. [1 mark]

Size = ______________ cm

13 Magnification can be calculated by dividing image size by the size of the real object.

$$\text{magnification} = \frac{\text{image size}}{\text{real size}}$$

Rearrange this equation to show how the real size of the object can be calculated. [1 mark]

14 The image below shows pollen grains viewed under a microscope. A special slide is used with a scale printed on it.

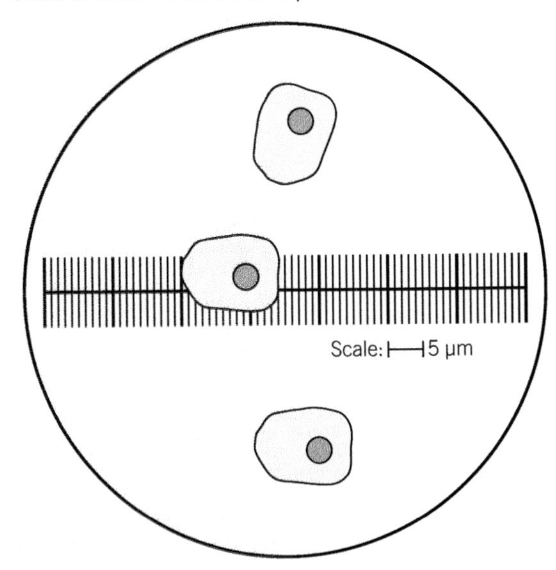

Use the scale to determine the length of a pollen grain. [1 mark]

____

____

Length of pollen grain = ____________ μm

15 **H** A student views a sample using a ×4 eyepiece and a ×100 objective lens. They measure it to be 2 cm long. Calculate the real size of the sample in μm. [3 marks]

____

____

____

Real size of sample = ____________ μm

16 **H** The nucleus of an animal cell has a diameter of 6 μm and the diameter of the whole animal cell is 100 μm. You can assume they are both perfect spheres. Calculate how many times larger the volume of the cell is compared to the volume of the nucleus. Give your answer as a whole number. [5 marks]

____

____

____

____

Answer = ____________ times larger

Exam Tip

You can be asked to apply maths in lots of different ways. You'll need to know equations from maths lessons in your science exam as well.

volume of a sphere $= \frac{4}{3}\pi r^3$

2 Osmosis

Investigate the effect of a range of concentrations of salt or sugar solutions on the mass of plant tissue.

Method

1. Use an apple corer to cut five vegetable cylinders and cut them to the same length (making sure that no skin is left on the end of the cylinder).
2. Dry each cylinder with filter paper before carefully measuring its mass.
3. Record these values in a table as 'Mass before'.

Concentration of solution in mol/dm^3	Mass before in g	Mass after in g	Percentage change in mass

4. Measure 10 dm^3 of each concentration of sugar solution into clearly labelled boiling tubes (distilled water should be used in a tube labelled 0 mol/dm^3).
5. Add one vegetable cylinder to each boiling tube and start the stop-clock.
6. After an exact amount of time (established by a preliminary experiment), remove the vegetable cylinders from the sugar solution.
7. Immediately dry each vegetable cylinder and re-weigh it.
8. Record these values in the table as 'Mass after'.
9. Calculate the percentage change in mass for each cylinder.

$$\% \text{ change in mass} = \frac{\text{change in mass}}{\text{initial mass}} \times 100$$

Safety

- Be very careful when using the sharp knife and the apple corer.
- Do not eat the plant/vegetable tissue samples.

Equipment

- plant tissue, e.g., potato, sweet potato, or beetroot
- range of sugar concentrations (or you could use different salt concentrations)
- distilled (pure) water
- apple corer
- sharp knife
- white tile
- filter paper
- tweezers
- boiling tubes
- measuring cylinder
- ruler
- balance
- stop-clock

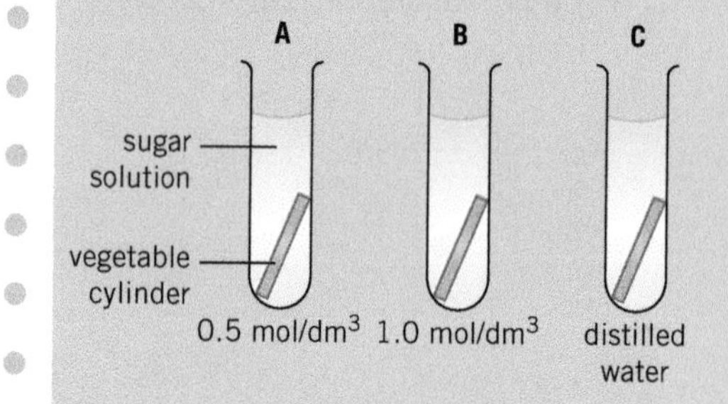

Remember

This practical requires you to accurately measure length, mass, and volume in order to measure osmosis. You should be able to describe how to do this for a range of sample types as osmosis happens in all cells.

The easiest example to use in class is a potato because they are cheap, readily available, and easy to cut into even shapes, but the principles of this experiment can be applied to any cell. There will be a range of examples used in this section to get you used to seeing the same theory applied with different types of tissues.

Exam Tip 

There are lots of opportunities for the exam board to test your maths skills in this topic. Don't get flustered by maths appearing in a biology question. Approach the maths questions the same way you would in a maths lesson.

1 Use answers from the box to complete the definition of osmosis. [1 mark]

concentrated	dilute	ions	water

The diffusion of water from a ________________ solution to a ________________ solution across a partially permeable membrane.

2 Give the meanings of the following keywords. [3 marks]

Hypertonic ________________________________

____

____

Hypotonic ________________________________

____

____

Isotonic ________________________________

____

____

Exam Tip

There are quite a few new words in this topic. It is important to learn 'exam perfect' definitions of keywords as you will often be asked to give the meaning of them for one or two marks.

3 Identify the part of the cell that is involved in osmosis.

Tick **one** box. [1 mark]

Cell membrane ☐

Chloroplast ☐

Nucleus ☐

Ribosome ☐

4 Describe what happens to a plant cell that is placed in a hypotonic solution. [6 marks]

____

____

____

____

____

____

____

____

____

____

Exam Tip

Long answer questions are the place to show the examiner that you know lots of keywords, but make sure you use them correctly!

5 Two students carried out a version of the experiment. They were given whole potatoes and were told to core and slice them into even cylinders.

Student A insisted that they should leave the skin on the end of the potato cores.

Student B thought that they should peel the potatoes before coring them.

Explain which student is correct. [3 marks]

Correct student = ____________________

Explanation:

____

____

____

____

____

6 Identify the independent variable in the experiment described in the method. [1 mark]

Independent variable = ____

7 A similar experiment investigated osmosis in carrots.

Explain why the carrot cores all needed to be the same length and width. [3 marks]

____

____

____

____

8 When this experiment was carried out with samples of chicken breast instead of vegetable cores, the following data was collected.

Initial mass in g	Final mass in g	Change in mass in g	Percentage change in mass
0.92	1.35		

Calculate the percentage change in mass of the chicken breast sample. [2 marks]

____

____

____

Percentage change in mass = ____________________ %

9 The following data was collected when the experiment was performed in class.

Concentration of sugar solution in mol/dm³	1	0.75	0.5	0.25	distilled water
Percentage change in mass	−39.65	−33.54	−29.34	−21.76	+12.63

a • Plot the data on the axes provided below.

• Draw a line of best fit. [3 marks]

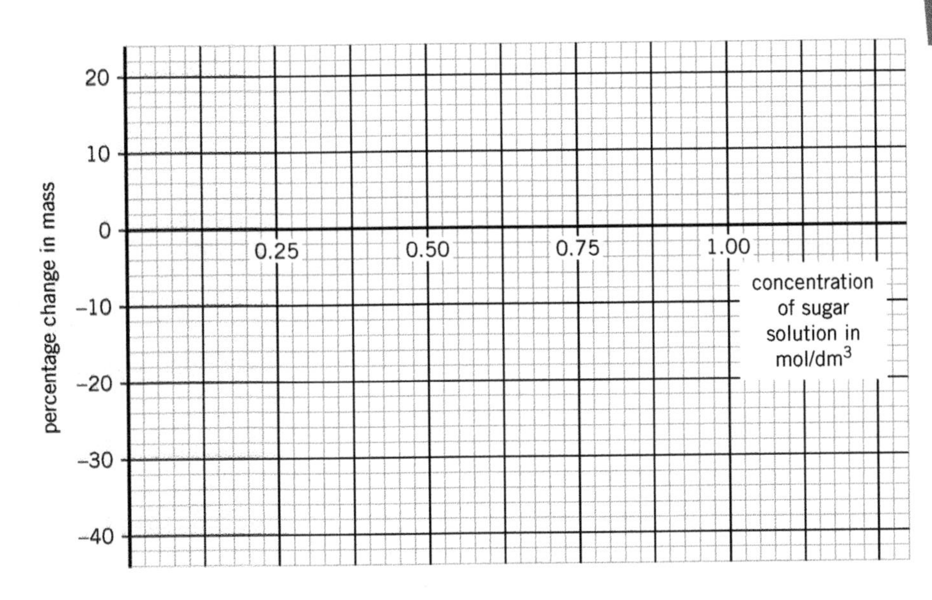

Exam Tip

Make sure you draw one smooth, continuous line of best fit. This is not art and the line should not be shaded or feathered.

b Estimate the concentration at which the sugar solution will be isotonic with the potato cells. [1 mark]

Concentration = ______________ mol/dm³

10 A teacher carried out preliminary experiments to find the best values for some of the control variables in this experiment.

- Identify two control variables that the teacher might want to determine in the preliminary experiment.
- Suggest a reason why it is important to use a suitable value for each of the variables you have chosen. [4 marks]

Control variable 1: ______________

Reason: ______________

____

____

Control variable 2: ______________

Reason: ______________

____

____

Hint

A preliminary experiment is carried out to a make sure the practical can be carried out within the time available. Think about what factors might have made the practical take too long.

11 A student placed an egg in vinegar to dissolve the shell, leaving the membrane exposed. The egg was then:

- dried, weighed, and placed it in 5% salt solution for 30 minutes
- removed from the solution, dried, and re-weighed.

The weight of the egg had not changed.

The student measured carefully and did not make any mistakes.

Explain what conclusion the student can draw from this experiment. [2 marks]

____

____

____

____

12 Two groups of students carried out a version of the experiment in exactly the same way but produced very different results.

Evaluate the methods the two groups used. [6 marks]

Group A:

- used one potato for the whole experiment
- weighed each individual potato core before placing in solution
- timed exactly 5 minutes before removing the samples
- weighed the cores immediately after removing from the solution.

Group B:

- took each core from a different potato
- weighed one core before and used this as the 'before' weight for all samples
- didn't time the experiment
- dried the samples before weighing them.

____

____

____

____

____

____

____

____

____

____

____

____

Exam Tip

When you see the command word evaluate you need to find good things, find bad things, give your opinion, and state why you came to that opinion.

13 Explain what will happen to the volume of water inside the visking tubing bag shown in the diagram below. [3 marks]

Change in volume of water inside the tubing bag:

____

Explanation:

____

____

____

14 A student cuts a cylinder of parsnip so that it has a diameter of 1.2 cm and a height of 6.4 cm.

a Calculate the volume of the cylinder. [2 marks]

____

____

Volume of parsnip cylinder = ______________ cm^3

Hint

volume of a cylinder = $\pi r^2 h$

b (H) Calculate the surface area of the cylinder. Give your answer to 2 decimal places. [2 marks]

____

____

____

Surface area = ______________ cm^2

Exam Tip

The surface area of a cylinder is two circles (top and bottom) and all the way around (a rectangle). Surface area of a cylinder = $2\pi rh + 2(\pi r^2)$

15 (H) When collecting the sample of sugar solution needed for the experiment, it was noticed that there were lumps of sugar at the bottom of the flask.

Suggest what has happened and what affect this will have on the results of your experiment. [3 marks]

____

____

____

____

____

____

3 Food tests

Test food samples for a range of carbohydrates, lipids, and proteins.

Method

A Test for starch

1. Add a few drops of iodine solution to the food on the spotting tile.
2. Yellow–red iodine solution turns blue–black if starch is present.

B Test for sugar

1. Place a small amount of food in a test tube.
2. Add enough Benedict's solution to cover the food.
3. Place the test tube in a warm water bath for 10 minutes.
4. Blue Benedict's solution turns brick red on heating if a sugar such as glucose is present.

C Test for lipids (fat)

1. Place a small amount of food into a test tube.
2. Add a few drops of ethanol to the test tube.
3. Shake the test tube and leave for one minute.
4. Pour the solution into a test tube of water.
5. Ethanol added to a solution gives a cloudy white layer if a lipid is present.

D Test for protein

1. Place a small amount of food in a test tube.
2. Add 1 cm^3 of Biuret reagent. Alternatively, add 1 cm^3 of sodium hydroxide solution and then add a few drops of copper sulfate solution.
3. Blue Biuret reagent turns purple if protein is present.

Equipment

- small pieces of different foods (e.g. cheese, crisps, pasta, ham, bread, boiled sweets)
- test tubes and test-tube rack
- spotting tile
- iodine solution
- Benedict's solution
- Biuret reagent *or* dilute sodium hydroxide solution and copper sulfate solution
- disposable pipettes
- filter paper
- water bath

Safety

- Do not eat any of the food.
- Some people may have food allergies.
- Wear splash-proof eye protection.
- Biuret reagent: IRRITANT.
- Sodium hydroxide: IRRITANT.
- Ethanol: HIGHLY FLAMMABLE – keep away from naked flames.
- Iodine solution: HARMFUL – avoid contact with skin.
- Water in the water bath will be very hot.

Remember !

This practical is testing for four different compounds found in food; lipids (fats), sugars, starch (carbohydrates), and proteins. You should be able to identify and describe the correct method for each test.

This practical also tests your knowledge of how to safely use heating devices and techniques.

Exam Tip

Make sure you learn the names of all the food tests and reagents. In an exam you may give the right colour change for a test but you will not get credit for your answer if you give the wrong test name. Be careful when the names of tests are similar, such as the Biuret reagent test for proteins and Benedict's test for sugars!

1 Give **two** advantages of knowing what compounds are in a food. [2 marks]

____

____

____

____

2 Give the type of enzyme in the digestive system that is responsible for breaking down the following substances in food. [3 marks]

a Carbohydrates

____

b Lipids

____

c Proteins

____

3 Draw **one** line from each food group to the products when it is broken down. [3 marks]

Food group	Products
Carbohydrates	Amino acids
Lipids	Fatty acids and glycerol
Proteins	Sugars

4 A sample was sent to a lab for testing, but the label fell off. Use the information given to suggest which substance the food contained. [1 mark]

Lab book entry — 2nd August

Test 1 – iodine test	result – negative
Test 2 – Biuret test	result – negative
Test 3 – lipid test	result – negative
Test 4 – Benedict's test	result – positive

Substance = ______________

5 A student tested a sample of food to see what compounds it contained. They added the following ingredients into the test tube.

- sample of food dissolved in 7.0 cm³ distilled water
- 1.5 cm³ of Biuret solution A
- 2.0 cm³ of Biuret solution B.

Calculate the percentage of the final solution that was Biuret solution B.
Give your answer as a whole number. [2 marks]

____

____

____

Percentage of final volume that was Biuret solution B = ______________%

6 It is important to select the correct equipment when carrying out an experiment.

Identify which of the following measuring cylinders would be the most appropriate for measuring 1 cm³ Biuret solution.
Give a reason for your answer.

[2 marks]

A 5 ml measuring cylinder

B 10 ml measuring cylinder

Most appropriate cylinder: ______________

Reason: ____

____

____

7 Read the following statements about the results of food tests.

A The starch test goes blue-black with a positive result.

B The lipid test goes purple if positive.

C The Biuret reagents test for proteins needs to be heated.

Tick **one** box to indicate which statements are true. [1 mark]

A, B, and C are true ☐

A only is true ☐

A and B are true ☐

B and C are true ☐

Exam Tip

There are four different mini practicals involved in this experiment. It is important that you learn and can recognise the results for each food test.

8 Safety information for the chemicals used in this practical is given below.

Chemical	Safety information
starch	can stain skin
ethanol	flammable
copper sulfate in Benedict's solution	irritant
sodium hydroxide in Biuret reagents	corrosive

Use the information in the table and your knowledge to describe the hazards, the risks associated with them, and how the risks can be managed. [6 marks]

Hint

Explain WHAT can hurt you, HOW it can hurt you, and how you can PREVENT it from hurting you.

9 Some processed cheeses have starch added to change the texture and to make them easier to melt and grate.

Suggest how this might affect the results of food tests on these cheeses. [2 marks]

10 Coeliac disease is a condition that means a person cannot eat foods which contain the protein gluten.

Describe how you would test for the presence of gluten in bread. [3 marks]

Hint

Don't be thrown by the fact you may not know much about gluten. Look carefully at what the question tells you about gluten.

11 Ethanol and bile, in the digestive system, have the same effect on lipids.

Choose the term that describes the effect of ethanol on lipids and give the meaning of the term. [3 marks]

A Denature

B Digest

C Emulsify

D Absorb

Correct term: ____________________

Meaning:

____

____

Hint

Bile has two functions. This question is not asking about bile's role in neutralising stomach acid.

12 Two students carried out Benedict's test for sugars.

Student A observed and recorded the colour change after 5 minutes.

Student B recorded data for 5 minutes using a colour probe (colorimeter) attached to a data logger. This constantly recorded the colour of the solution as a numerical value.

a Evaluate these two different methods of following the practical. [6 marks]

____

____

____

____

____

____

____

____

____

____

____

____

Exam Tip

When you see 'evaluate' as the command word in an exam question, you need to point out the good bits and the bad bits of both methods, give your opinion about which is best, and then explain why you came to that conclusion.

b Student A's observation was a qualitative result, whereas student B's results were quantitative.

Describe what the terms 'qualitative' and 'quantitative' mean. [2 marks]

Qualitative

____

____

Quantitative

____

____

13 Colorimeters can be used to measure the amount of red light absorbed by a sample with Benedict's reagent in it.

A calibration curve can be created using a set of solutions with known concentrations of glucose.

a To complete the calibration curve:

- plot the data from the table below
- draw a line of best fit. [3 marks]

% concentration of sugar solution	0	2	4	6	8	10
% absorption of red light	1.9	1.0	0.48	0.25	0.1	0.02

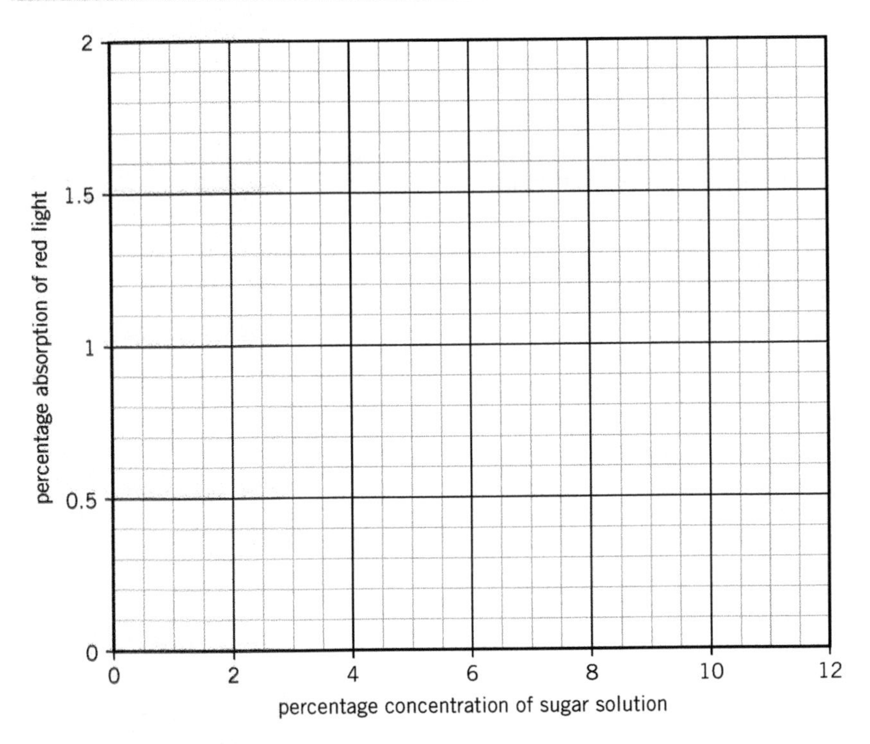

b A student tested a range of unknown samples (A–D) for sugar using Benedict's reagent and the calibrated colorimeter.

Use the student's results (in the table below) and the calibration curve to estimate the percentage of sugar in each solution. [4 marks]

Sample	% absorption of red light	Estimated % concentration of sugar solution
A	1.2	
B	0.7	
C	0.9	
D	1.5	

4 Enzymes

Investigate the effect of pH on the rate of reaction of an enzyme.

Method

1. Transfer 2 cm^3 of each pH buffer solution to separate, labelled test tubes. Use a separate syringe for each pH buffer.
2. Use another syringe to add 4 cm^3 starch solution to five test tubes.
3. Place the pH buffer test tubes, starch solution test tubes, and a test tube containing 10 cm^3 amylase solution in a 30°C water bath.
4. Place a thermometer in one of the test tubes containing the starch solution and wait until it reaches 30°C.
5. Whilst waiting, add a drop of iodine solution into each dimple of a spotting tile.
6. Use a glass stirring rod to transfer a drop of starch solution to the first dimple of the spotting tile. This will be your 'zero time' test.
7. When the solutions have reached 30°C, add 2 cm^3 of the first pH buffer solution 2 cm^3 amylase solution to one of the starch solution test tubes and start a stop clock.
8. Every 10 seconds, use the stirring rod to transfer a drop of the mixed solution to the iodine solution in the next dimple on the spotting tile. Make sure the stirring rod is rinsed with water in between each sample.
9. Repeat step 8 until the iodine in the dimples does not change colour.
10. Record the time for amylase taken for amylase to completely break down the starch in a suitable results table.
11. Repeat steps 7–10 for each pH buffer solution.

Safety

- Use eye protection.
- Iodine is harmful. Avoid contact with skin.

Equipment

- amylase solution (0.5%)
- starch solution (0.5%)
- iodine solution in a dropper bottle
- buffer solutions (range of pH values)
- 5 cm^3 syringes
- pipette
- test-tube rack
- test tubes (one for each pH to be tested)
- spotting tile
- stop clock
- marker pen
- water baths set at 30°C

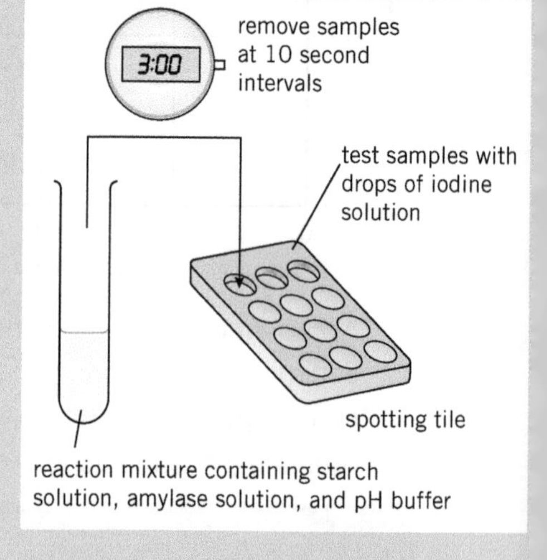

Remember

This practical tests your ability to accurately measure and record time, temperature, volume, and pH. You should know how to find the rate of a reaction by measuring the time taken for an indicator to change colour. You should be able to describe how to use a 'continuous sampling' technique to monitor a reaction. Make sure you can explain how a water bath is used to control the temperature and why this is important.

Exam Tip 

The enzyme used in this experiment is amylase, but the same principles can apply to any enzyme and its substrate.

1 Define the term enzyme. [2 marks]

____

____

____

____

Exam Tip

There are lots of key words in Biology. These are often asked as one- or two-mark questions. It is worth spending the time to learn as many as you can.

2 The enzyme used in this practical is amylase.

Give the sites where amylase is produced in the body. [2 marks]

____

____

Exam Tip

Always look at the number of marks available for a question. This gives you a good indication of how many points you need to write down. This question is worth two marks, so you need to give two places where amylase is produced.

3 Describe the roles of the following solutions that are used in this practical. [3 marks]

Starch solution

____

____

Iodine solution

____

____

____

4 Suggest why it is important to have a drop of iodine in the spotting tile as a 'zero time'. [1 mark]

____

____

5 This experiment tests how pH affects the rate of an enzyme-controlled reaction. Identify two other factors that would affect the rate of an enzyme-controlled reaction. [2 marks]

____

____

6 Catalase is a substance found in the liver that helps to break down hydrogen peroxide into water and oxygen.

Use the 'lock and key' model to describe how catalase breaks down hydrogen peroxide into water and oxygen. [6 marks]

Hint

Any word that ends in -ase is a type of enzyme.

7 **a** Explain why it is important to allow all the solutions to reach the temperature of the water bath before mixing the starch solution and amylase solution. [2 marks]

b It is best to monitor the temperature by placing a thermometer in one of the test tubes containing the starch solution, rather than placing it directly in the water bath.

Suggest why the thermometer is placed in the tube and not in the water bath. [2 marks]

8 Some bacteria live in volcanic vents where the temperature is 95°C. These bacteria can survive because their enzymes are specialised to work well at high temperatures.

On the axes below, sketch and label two lines to represent reactions controlled by:

- a human enzyme
- an enzyme from a volcanic vent bacterium. [4 marks]

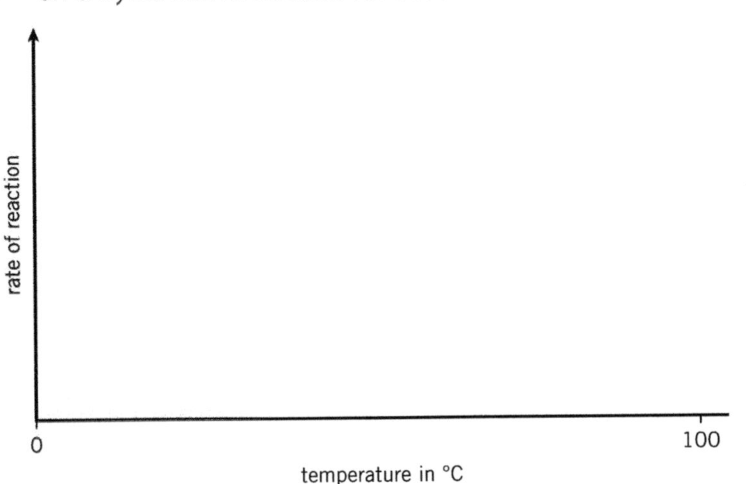

Exam Tip

When you're asked to sketch a graph, you need to include key details, for example, axes labels (if not already provided) and a line showing the correct shape of the graph, or the correct trend. You don't need to add number labels on the axes or to plot individual points unless you are specifically asked to do so.

9 The graph below shows the rate of reaction for an enzyme-controlled reaction at pH values from 0 to 4.

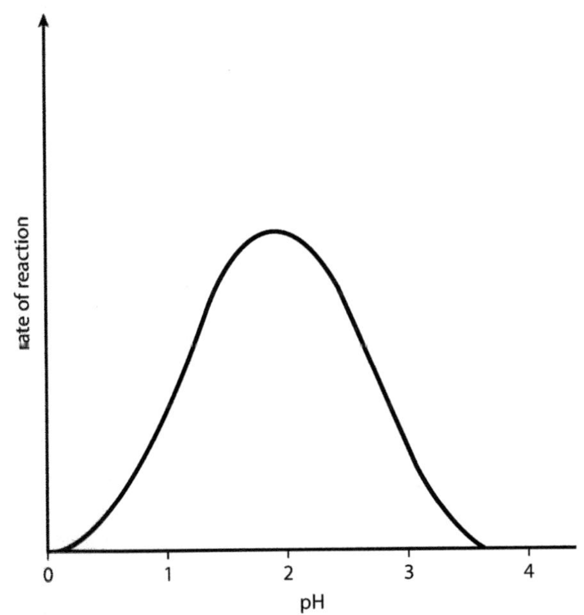

Explain the shape of the graph. [3 marks]

10 Three different classes carried out a version of this experiment.

Fresh amylase solution was made up for class A on Monday morning.

Class B used the same amylase solution at the end of Monday.

Fresh amylase solution was then made up for class C on Wednesday morning.

a Plot the results in the table below on the axes provided. [6 marks]

pH of solution	Average time for enzyme to break down substrate in s		
	Class A	Class B	Class C
4	43	72	35
5	25	52	19
6	57	198	37
7	170	340	120

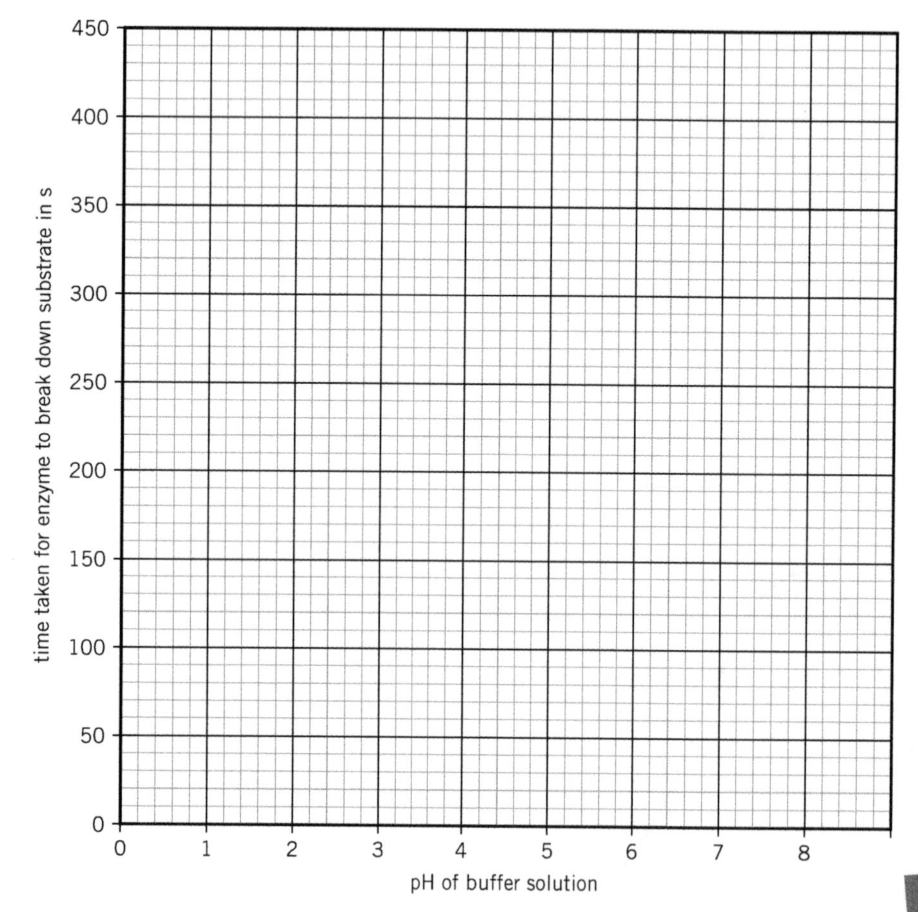

b Suggest reasons for the differences between the results of each class. [2 marks]

____

____

____

____

Hint

This graph may look unusual or upside-down, don't let that confuse you. Just read the axes carefully and think logically about what the data is saying.

c
- Estimate the optimal pH for the enzyme used.
- Give evidence from your graph to support your estimate. [2 marks]

____

____

____

____

11 A student carried out this experiment but found that the reaction happened so quickly that even at 10 seconds there was no colour change.

Suggest how the student could change their method to make the results easier to measure. [2 marks]

____

____

____

____

12 Describe why it is important to rinse the glass rod with water between each sample. [1 mark]

____

____

____

____

5 Photosynthesis

Investigate the effect of light intensity on the rate of photosynthesis.

Method

1. Cut a 10 cm piece of pondweed.
2. Place the piece of pondweed into a beaker of water, covered with an inverted filter funnel. Make sure the cut end of the pondweed is at the top.
3. Fill a measuring cylinder with water and carefully invert it over the top of the filter funnel.
4. Position a lamp exactly 100 cm from the pondweed. Switch the lamp on and leave it for two minutes to allow the pondweed to acclimatise.
5. Start a stopclock and record the number of bubbles produced in three minutes.
6. After 3 minutes, record the volume of gas that has been collected in the measuring cylinder.
7. Refill the measuring cylinder with water and repeat steps 4–6 for distances of 80 cm, 60 cm, 40 cm, and 20 cm.

Equipment

- pondweed
- scissors
- lamp
- large beaker
- filter funnel
- 10 cm^3 measuring cylinder
- metre ruler
- stop clock

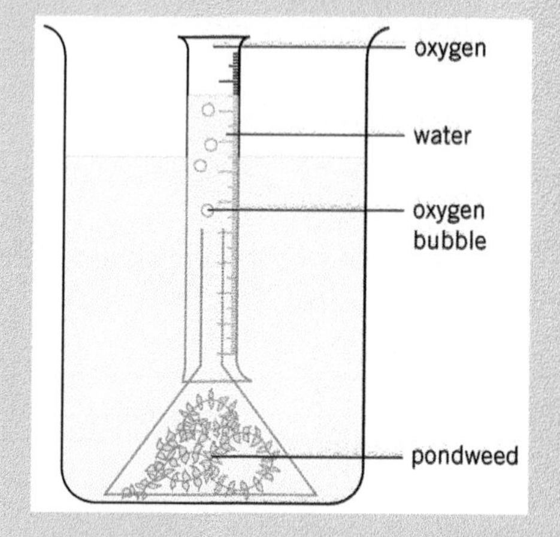

Safety

- Wash hands after contact with pondweed and pond water.
- Lamp may get hot.
- Keep electrical equipment dry and do not handle if hands are wet.
- Dispose of the pondweed responsibly.

Remember !

The skill being tested in this practical is your ability to accurately measure changes in the rate of photosynthesis in response to changes in the environment. You should be able to describe how to measure the rate of a reaction or biological process by collecting a gas that is produced.

Remember that you will probably have plotted a graph of the volume of gas collected (or number of bubbles) against the distance between the lamp and the pondweed (or light intensity). It is important to know what this graph looks like and to be able to explain its shape.

Exam Tip

- A living plant is needed for this experiment to work properly. Plants only undergo photosynthesis in the right conditions, and a number of factors need to be controlled. The following questions will change some of those factors, some in an obvious way and some in a not so obvious way.
- You need to know the word equation for photosynthesis. You should also be able to recognise the chemical formulae for all the substances in the equation.

1 Identify the dependent variable in this experiment. [1 mark]

____

____

2 Draw one line from each compound to its chemical formula. [3 marks]

carbon dioxide	$C_6H_{12}O_6$
glucose	H_2O
oxygen	CO_2
water	O_2

3 Complete the word equation for photosynthesis. [4 marks]

__________ + __________ $\xrightarrow{\text{Light}}$ __________ + __________

4 Identify the gas in the bubbles produced at the cut end of the pondweed.

[1 mark]

____

5 Two students carried out this practical with different light sources.

- Student A used a desk lamp.
- Student B used the light on their mobile phone.

Suggest how you would expect the students' results to differ and explain why.

[3 marks]

____

____

____

____

6 Two groups of students carried out the experiment in exactly the same way, their results are show below.

Group A

Distance to light source in cm	10	20
Number of gas bubbles	120	74
Volume of gas collected in cm^3	19	13

Group B

Distance to light source in cm	10	20
Number of gas bubbles	40	27
Volume of gas collected in cm^3	18	12

Compare the two sets of results. [4 marks]

Exam Tip

When you have 'compare' as the command word, you need to mention the similarities and the differences in your answer.

7 A student has samples of two different species of pondweed, shown below.

Explain whether it would be a fair test to compare rates of photosynthesis between these two samples of pondweed. [3 marks]

8 The following results were obtained after repeating the experiment. Complete the table.

Distance from light source in cm	Number of gas bubbles released in one minute			
	Test 1	Test 2	Test 3	Mean
10	40	42	38	
20	27	25	68	
30	15	17	19	
40	11	12	9	

Exam Tip

Whenever you are asked to calculate the mean in a table of results, you should check for anomalous results first as there will often be one hiding in there. Do not include this result when working out the mean.

9 A student repeated the experiment twice in a day. The sample of pondweed was:

- tested at the start of the day
- left in water in the sun for the duration of the day
- retested at the end of the day.

Suggest why the volume of gas collected at the end of the day was much lower than the volume of gas collected at the start of the day. [4 marks]

____

____

____

____

____

____

10 Give the name of the structure in plant cells where photosynthesis takes place. [1 mark]

____

11 A student sets up their apparatus with air already trapped in the measuring cylinder before starting the experiment.

a Identify the type of error that the student has introduced. [1 mark]

Anomalous error ☐

Human error ☐

Random error ☐

Systematic error ☐

b Describe how the results would be adjusted to compensate for this error. [1 mark]

____

____

12 Suggest an alternative to a measuring cylinder that could be used to accurately measure the volume of gas collected. [1 mark]

____

____

13 A student investigates the effect of light intensity on the rate of photosynthesis. Their results are shown in the graph below.

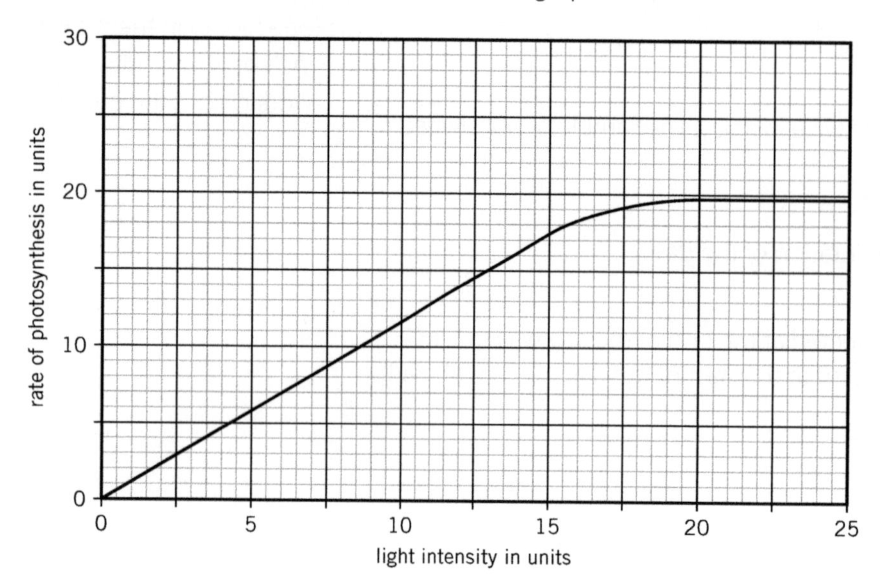

a Give the rate of photosynthesis when the light intensity is 12.5 units. [1 mark]

Rate of photosynthesis = ____________ units

b Increasing the light intensity beyond 20 units will not lead to a greater increase in the rate of photosynthesis.

Give evidence from the graph that supports this statement and explain why this is the case. [3 marks]

Evidence ______________________________________

Explanation ____________________________________

____

____

Exam Tip

Make sure you draw construction lines on the graph. These will help you work out the answer and will show the examiner you know what you're doing.

14 Describe possible sources of error in this experiment and suggest how the experiment could be adapted to reduce them. [6 marks]

15 (H) As the distance between the lamp and the pondweed increases, the light intensity decreases according to the inverse square law.

Explain the inverse square law. [4 marks]

Exam Tip

You might find it helpful to include a diagram in your answer – but don't spend too long drawing it!

6 Reaction time

Plan and carry out an experiment to investigate how one factor affects human reaction time.

Method

Human reaction times can be affected by several factors. You are going to choose one factor to investigate.

You will write down a hypothesis about how you think reaction time will change as you change your factor.

The next thing you will do is to test your hypothesis.

1. Working in pairs, one student holds a ruler vertically so that zero is at the bottom. The other student rests their hand on the edge of the bench and puts their thumb and fingers in a 'C' shape around the ruler, level with the zero marking.
2. Without warning, the first student drops the ruler and the second student has to catch it as quickly as they can.
3. Write down the number just above the second student's thumb. The lower the number, the faster the reaction time. Write the number in the results table.
4. Repeat steps 1–3 another four times.
5. Change the factor that you have decided to investigate (e.g., caffeine consumption, being distracted, gender) and repeat steps 1–4.
6. Use a conversion table to convert the results into reaction times.
7. Calculate mean values for each set of five results and plot your mean results on a bar chart.

Equipment

- metre ruler
- reaction time conversion table
- caffeinated soft drink (if caffeine consumption is the factor being changed)

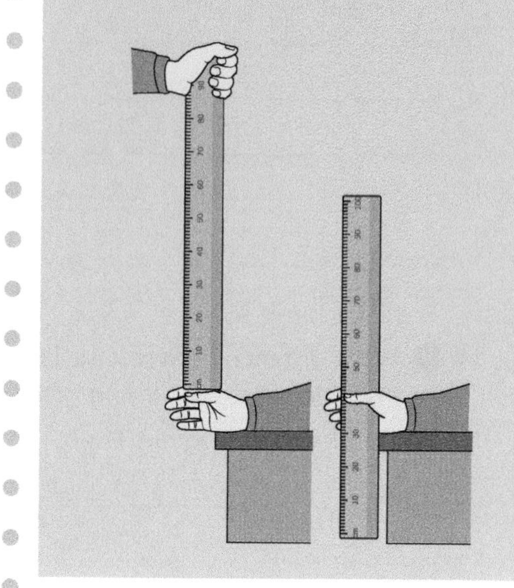

Safety

- Do not consume drinks in the laboratory.

Remember

This practical tests your ability to plan an experiment and choose suitable variables to change in your investigation. You should be able to write a hypothesis predicting the effect that changing one variable will have on your results. Make sure you know what independent and dependent variables are.

Exam Tip 

You don't need to be able to calculate reaction time from the ruler measurement. In the practical you are given a table to look this data up, so make sure you know how to use the table.

1 **'Reaction time is affected by how tired a person is.'**

Describe how you would change the independent variable to test this hypothesis.

[3 marks]

____

____

____

____

____

____

____

Hint

The number of marks can give you a clue about how much detail to include. There are only three marks here so you are not going to be expected to re-write the whole method in detail.

Focus on the details of how you would change the factor.

2 A student wants to test the hypothesis:

'Reaction time is affected by taking illegal drugs.'

Explain whether this would be a suitable hypothesis to test.

[2 marks]

____

____

____

____

____

3 Draw one line from each factor to the effect it will have on reaction time.

[3 marks]

Factor being changed	Effect on reaction time
Caffeine	Decrease reaction time
Alcohol	Increase reaction time
Being distracted	No effect

Hint

Pay close attention to the wording of the effects and don't confuse reaction speed with reaction time, i.e., faster reactions = decrease in reaction time.

4 Three students recorded the following results for the ruler drop experiment.

Drop	Ruler measurement in cm		
	Student A	Student B	Student C
1	48.0	40	38.00
2	46.0	39	46.00
3	51.0	41	43.00
4	46.0	42	39.00
5	43.0	43	38.00
mean			
range			

a Complete the table by calculating the range and mean of each set of results. [5 marks]

b Give the letter of the student whose results are the most precise and give a reason for your answer. [2 marks]

Letter ____________________

Reason ____________________

____

c Use the following formula to calculate which student's results have the greatest percentage uncertainty. [4 marks]

$$\text{percentage uncertainty} = 100 \times \frac{(\text{range} \div 2)}{\text{mean}}$$

____

____

____

Student with greatest percentage uncertainty = ____________________

d Explain the difference between precise results and repeatable results.

[2 marks]

____

____

____

____

5 Two students are investigating whether listening to music has an effect on reaction time.

a Explain why it is important that they use the same hand to catch the ruler each time. [2 marks]

____

____

____

Exam Tip

If you are being asked about why a variable should be kept the same, it is usually worth starting your answer with 'To make it a fair test'.

b State two other variables which should be kept the same in the students' experiment. [2 marks]

Variable 1 ________________________________

Variable 2 ________________________________

6 Describe what happens in a student's nervous system as they react to the falling ruler. [6 marks]

____

____

____

____

____

____

____

____

____

____

____

____

7 Two groups of students measured reaction times in different units.

One group recorded the data in mm and the other group recorded it in metres.

Give 0.28 m in mm. [1 mark]

Answer = ____________ mm

8 Two students used an alternative method using a stopwatch.

Student A started the stopwatch as they released the ruler.

When they saw student B catch the ruler, student A stopped the stopwatch.

They then recorded the reaction time directly from the stopwatch.

Evaluate this method. [4 marks]

____

____

____

____

____

____

Exam Tip

When you see the command word 'evaluate', you need to give points for and against the method. You should then use these points to justify your opinion about whether the method is a good idea or not.

9 Two students have a hypothesis that:

'Reaction times will become faster with more practice.'

To test this hypothesis they repeat the ruler drop ten times to see if there is any change.

a Plot their results shown in the table on the axes provided. [4 marks]

Drop attempt	Reaction time in seconds
1	0.32
2	0.31
3	0.30
4	0.28
5	0.17
6	0.27
7	0.26
8	0.24
9	0.23
10	0.22

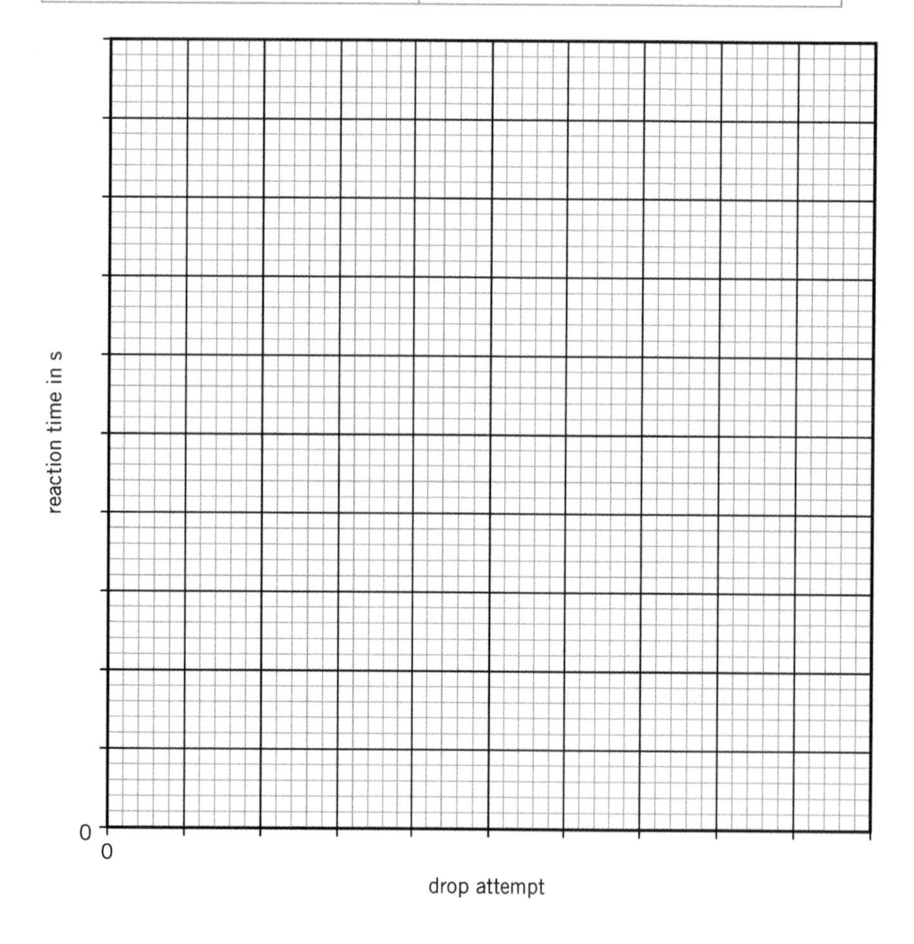

Exam Tip

When there is no scale provided, you will have to label your own axes. You will be given a sensible size of graph paper, so if your scale doesn't use most of it, check that you're going up in sensible amounts. If your scale uses less than half the space then you can double it!

b Identify which result was an anomaly. [1 mark]

Anomalous drop number = ____________________

c Describe the pattern that can be seen in the results. [2 marks]

___

___

d Explain whether the results support the students' original hypothesis. [2 marks]

___

___

___

e The students then repeated the all ten drops while the 'catcher' was watching TV at the same time.

On your graph in part **a**, sketch a line of the results you would predict for this experiment. [2 marks]

7 Field investigations

Investigate the population size of a plant species in a habitat.

Method

A Transect line

1. Stretch a 20 m tape measure from the base of a tree to an open area of ground.
2. Place the quadrat at exactly 2 m on the tape measure.
3. Count and record the number of plants of the species being investigated that fall within the quadrat.
4. Record the light intensity at this point on the transect line.
5. Repeat measurements every 2 m along the tape measure.

B Random sampling

1. Place two 20 m tape measures (labelled X and Y) at right angles to each other to form the sides of a 20 m^2 square area.
2. Put two sets of cards in a bag, each with the numbers 1 to 20 on them.
3. Pull two numbers out of the bag. The first number indicates how many metres along tape measure X you should move. The second tells you how far along tape measure Y to move.
4. Place the quadrat at these co-ordinates.
5. Count and record the number of plants of the species being investigated that fall within the quadrat.
6. Repeat steps 2–5 until you have sampled ten quadrats.

Safety

- Follow local rules on working in an outside environment and wash hands after the lesson.
- When any fieldwork is undertaken, work in groups and be aware of any hazards in that specific environment.
- Sensible footwear and clothing should be worn. If the weather is hot and sunny, sunscreen and hats are required.

Equipment

- two 20 m measuring tapes
- two sets of 20 cards, each numbered from 1 to 20
- 50 × 50 cm gridded quadrat frame
- notebook and pencil
- identification sheet
- optional equipment to measure abiotic factors, such as light meter, pH meter/ universal indicator paper, anemometer

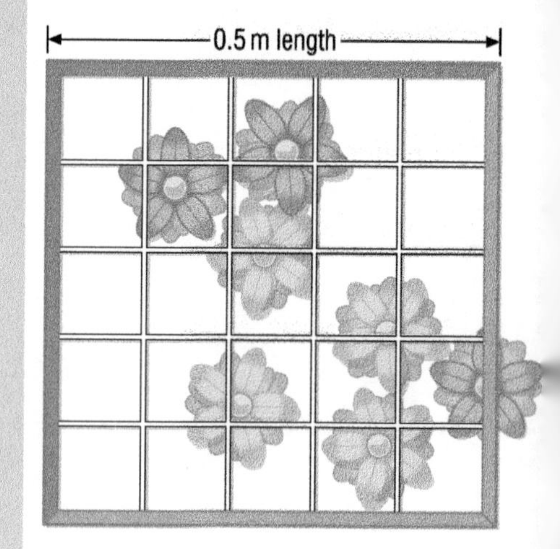

Remember !

This practical tests your ability to accurately measure length and area and to apply appropriate sampling techniques in the field. There are two different methods of sampling with quadrats covered by this practical. It is important that you can describe how to do both of them and describe what the purpose of each method is.

Exam Tip

You should know how to calculate an estimate of the total population of a species using the equation:

$$\text{estimated population size} = \frac{\text{total area}}{\text{area sampled}} \times \text{number of individuals counted}$$

1 Why it is important to use a system to generate random coordinates instead of just choosing 'random' locations to place the quadrat.

Tick **one** box [1 mark]

Avoids unconscious bias ☐

Increases sample size ☐

It is faster ☐

Reduces errors ☐

2 Define 'habitat'. [1 mark]

____

____

3 A company wants to build a new headquarters on a piece of grassland that has an area of 21 000 m^2. They have been told they must carry out an environmental assessment before building begins.

The environmental assessment is carried out by randomly sampling ten 1 m^2 areas within the grassland.

a Suggest why it is important to investigate the species in the grassland, and to estimate their population sizes and distributions. [2 marks]

____

____

____

____

b Give an advantage of estimating population size instead of measuring the true population size in the grassland. [1 mark]

____

____

c Suggest an improvement to the method that would improve the accuracy of the population size estimate. [1 mark]

____

____

4 The graph below shows measurements of soil pH and the % grass coverage along a transect line in a garden. Soil pH has been plotted.

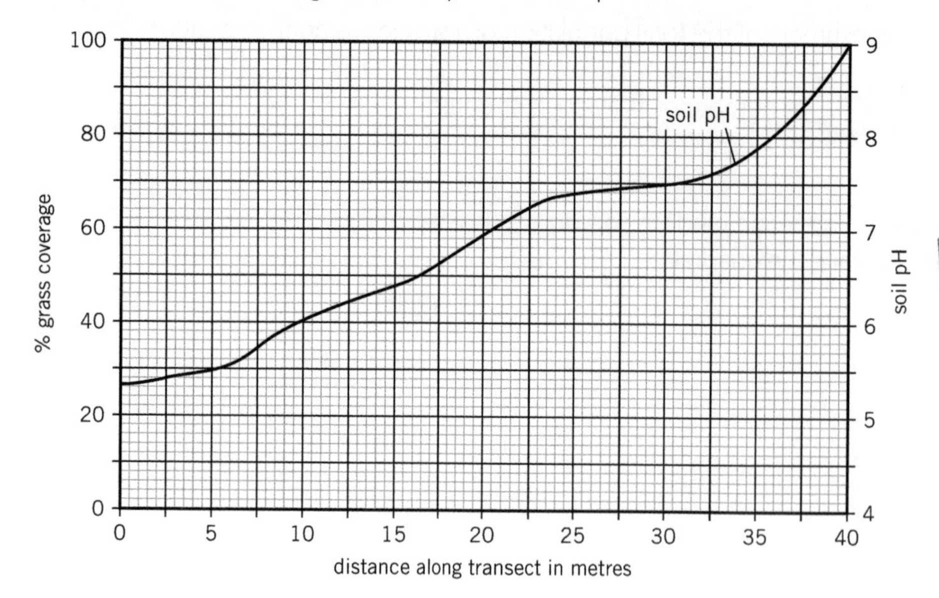

a Give the pH of the soil 20 metres along the transect line. [1 mark]

pH of soil = ___________

b Plot the following data on the graph. [2 marks]

Distance along transect line in m	% grass coverage
5	50
10	85
15	95
20	90
25	35
30	40
35	30
40	15

c A gardening advice website says:

'Grass grows best in soil with a pH between 6 and 6.8.'

Explain whether the graph supports this statement. [2 marks]

____

____

____

____

5 Give 1 m^2 in cm^2. [1 mark]

____

____

1 m^2 = ___________ cm^2

Exam Tip

Be careful when graphs are shown with more than one y-axis. Make sure you read the correct axis. In this case, the soil pH is shown on the right.

Hint

A common mistake is to write 100 cm^2 – this is not the correct answer.

6 A student investigates the distribution of plantain weeds along a transect line and writes a list of factors they think might affect the distribution of the weeds.

Complete the table by sorting the student's list into biotic and abiotic factors.

[2 marks]

nutrient levels	light intensity	parasites	soil pH
primary consumers	temperature	wind	

Biotic factors	Abiotic factors
competition from other plants	moisture levels

7 Describe a method to estimate the total population of ash trees in a 10 km^2 section of forest. [6 marks]

8 Random sampling and transect lines have different purposes.
Describe the differences in what the two methods are used to investigate.

[4 marks]

9 Calculate the area of a 25 cm × 25 cm quadrat in m^2. [2 marks]

Area = ________ m^2

10 A student investigated the distribution of two plant species along a transect line. The transect started next to a wall and finished in the middle of a sunny school field.

Quadrat number	Distance from wall in m	Number of plant A	Number of plant B
1	0	0	10
2	1	0	10
3	2	1	11
4	3	0	7
5	5	1	5
6	9	0	4
7	15	1	1
8	17	2	0
9	19	3	0
10	25	5	0

a Compare the results for the two plant species. [3 marks]

b Look again at the data in the table.

Suggest a change to the student's method that would have improved the data. [1 mark]

c A student found that plants near the wall had larger leaves than the ones in the school field.

Explain this finding. [2 marks]

Exam Tip

The command word 'compare' means you need to describe the similarities and/or differences between things. Don't just describe each of them on their own.

11 A student used a 75 cm × 75 cm quadrat to randomly sample daisies in the school field. The field measured 30 m × 75 m.

Quadrat number	Number of daisies counted
1	12
2	21
3	13
4	89
5	24
6	19
7	10
8	29
9	21
10	16

a Calculate the mean and median number of daisies. [2 marks]

Hint

The median of an even set of numbers is the mean of the middle two numbers when the numbers are sorted into ascending order.

Mean = ______ daisies

Median = ______ daisies

b Use the data collected to estimate the population size. [5 marks]

Estimate of population size = ______ daisies

8 Making salts

Prepare a pure, dry sample of a salt by reacting an insoluble metal oxide or metal carbonate with a dilute acid.

Method

Record all your observations during this practical

1. Using a measuring cylinder, measure 20 cm^3 of dilute sulfuric acid into the beaker.
2. Add half a spatula of copper oxide into the dilute sulfuric acid and stir with the glass rod.
3. Warm the beaker and its contents gently. Do not allow the reacting mixture to boil.
4. Continue adding the copper oxide in small amounts until no more dissolves (this should be most of the solid you have been provided with).
5. Set up a filter funnel and filter paper in a conical flask. Filter the mixture and discard the unreacted copper oxide.
6. Pour the filtrate into an evaporating basin and place it on a beaker of water. Heat the water until the volume of the solution in the evaporating dish is halved.
7. Remove from the heat. When cool, stand the evaporating basin on a piece of paper with your name on it. Leave it to crystallise overnight.
8. During the next lesson, remove the crystals from the concentrated solution with a spatula and gently pat them dry between two pieces of filter paper.

Equipment

- 25 cm^3 measuring cylinder
- 100 cm^3 beaker
- spatula and glass rod
- filter paper and funnel
- Bunsen burner, heat proof mat, tripod, gauze, and tongs
- evaporating basin
- 1 mol/dm^3 sulfuric acid
- copper oxide powder

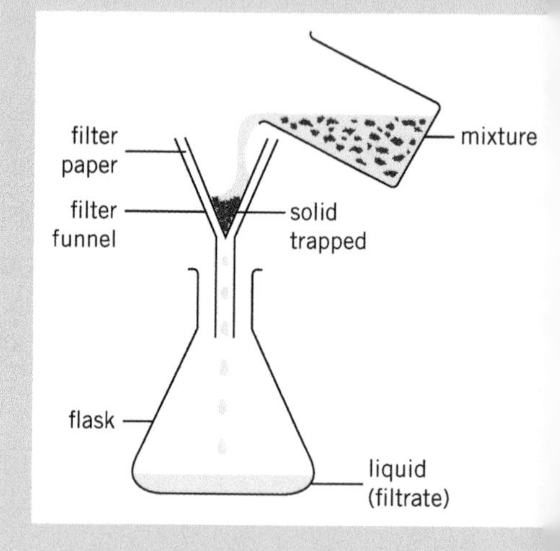

Safety

- Eye protection should be worn at all times.
- Do not allow the reacting mixture to boil.
- 1 mol/dm^3 sulfuric acid: IRRITANT
- 2 mol/dm^3 hydrochloric acid: IRRITANT
- copper oxide powder: HARMFUL

Remember !

This required practical is testing whether you can safely separate and purify a chemical mixture. You need to be able to describe how filtration, evaporation, and crystallisation can be used to make pure, dry samples of soluble salts. You should also be able to describe how substances can be tested for purity.

Exam Tip

You can be asked about any salt being produced, not just the examples you are familiar with. You need to learn all of the general acid equations and be able to apply them.

metal + acid → salt + hydrogen

metal oxide + acid → salt + water

metal hydroxide + acid → salt + water

metal carbonate + acid → salt + water + carbon dioxide

Using hydrochloric acid will lead to the production of chloride salts.

Using sulfuric acid will lead to the production of sulfate salts.

Using nitic acid will lead to the production of nitrate salts.

1 Identify the hazards in this practical, describe the risks associated with them, and suggest what can be done to prevent the risk happening. [6 marks]

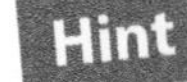

Hint

Think about WHAT can harm you, HOW it can harm you and how you can PREVENT it from harming you.

____

____

____

____

____

____

____

____

____

____

2 The crystals at the end are copper sulfate. The formula for copper sulfate is $CuSO_4$. Which of the following statements are true?

A $CuSO_4$ is comprised of 3 elements

B $CuSO_4$ is comprised of 4 elements

C $CuSO_4$ is comprised of two types of ion.

Tick **one** box. [1 mark]

A only ☐

B only ☐

B and **C** ☐

A and **C** ☐

Hint

- On the periodic table, the mass number is the larger of the two numbers beside an element's symbol.
- The small, subscript number in a formula tells you how many atoms of each element there are. If there is no small number after an element, that means there is just one of that element in the compound.

3 Calculate the relative formula mass of sulfuric acid, H_2SO_4. [1 mark]

____

____

Relative formula mass of H_2SO_4 = ____________

4 Suggest the function of the filter paper. [1 mark]

____

____

Hint

What was left in the filter paper once you were finished with it?

5 Draw one line from each key term to its definition. [2 marks]

Key term	Definition
Filtrate	An insoluble solid formed by a reaction taking place in a solution.
Precipitate	A solution that has passed through a filter.
Salt	A compound formed when the hydrogen in an acid is replaced by a metal.

6 The method described for this practical evaporates the water from the solution by putting the evaporating basin on a water bath.

Suggest **one** other method of evaporating the liquid. [1 mark]

____

____

7 Describe the purpose of evaporating the water. [1 mark]

____

____

8 A student carries out a similar experiment with copper oxide and sulfuric acid and makes two observations.

- The black powered appears to 'disappear'.
- The solution turns from colourless to blue.

Explain why these changes happened. [4 marks]

____

____

____

____

____

____

9 Complete and balance the symbol equation for the reaction of copper(II) oxide with sulfuric acid. [2 marks]

$$\_\_\_\_\_\_\_\_ + H_2SO_4 \rightarrow \_\_\_\_\_\_\_\_ + \_\_\_\_\_\_\_\_$$

10 The formula of copper(II) oxide is CuO.

Write the formula of copper(I) oxide. [1 mark]

____

____

11 Hydrochloric acid can be reacted with magnesium or magnesium carbonate. Both give magnesium chloride as one product and a gas as another product.

magnesium + hydrochloric acid → magnesium chloride + gas A

magnesium carbonate + hydrochloric acid → magnesium chloride + water + gas B

Identify gas A and gas B and give the test to confirm the identity of each gas. [4 marks]

Gas A is ______________

Test for gas A:

____

Gas B is ______________

Test for gas B:

____

12 A salt is a compound formed when an acid reacts with a base. Write word equations for the production of the following salts.

a The production of potassium chloride [2 marks]

____

b The production of iron sulfate [2 marks]

____

c The production of lead nitrate [2 marks]

____

Exam Tip

It is important that you can apply any of the general acid equations.

13 Insoluble calcium sulfate is produced in the reaction below.

$H_2SO_4(aq) + Ca(OH)_2(aq) \rightarrow CaSO_4(s) + 2H_2O(l)$

Explain what you would observe if you carried out this reaction. [2 marks]

____

____

Hint

Look carefully at the state symbols. Copper sulfate is a soluble salt, shown by the state symbol (aq). However, calcium sulfate is an insoluble salt and in the equation, we show this by using the state symbol (s).

14 Suggest why it is important that the copper oxide is in excess in this practical. [1 mark]

____

____

15 (H) Calculate how much copper sulfate can be produced from 17.0 g of copper oxide.

Give your answer to three significant figures. [3 marks]

____

____

____

____

____

Mass of copper sulfate produced = ______________ g

9 Electrolysis

Investigate the decomposition of two ionic solutions using electrolysis.

Method

1. Transfer 50 cm³ copper(II) chloride solution to the 100 cm³ beaker.
2. Insert the carbon electrodes into the solution, ensuring they do not touch.
3. Attach the electrodes to the DC power supply using the leads and crocodile clips.
4. Switch the power supply on at 4 V and carefully observe what happens at the anode and the cathode. Record your observations.
5. Position a piece of damp litmus paper above the solution next to the anode. Record your observations.
6. Collect a new set of electrodes, wash the equipment, and repeat steps 1–5 with 50 cm³ sodium chloride solution in the beaker.

Safety

- copper(II) chloride solution – IRRITANT
- oxygen gas – OXIDISING
- hydrogen gas – EXREMELY FLAMMABLE
- chlorine gas – TOXIC
- The electrolysis of brine produces a solution of sodium hydroxide, which is corrosive.
- Wear chemical splash proof eye protection.
- Wear nitrile gloves and only complete the practical in a well-ventilated room. Take extra care if you are asthmatic.
- Switch off the electric current as soon as you have made your observations.

Equipment

- 100 cm³ beaker
- 2 × carbon electrodes
- 2 × crocodile clips and wires
- 1 × low voltage lab pack
- copper(II) chloride solution
- saturated sodium chloride solution
- litmus paper
- forceps
- eye protection and nitrile gloves

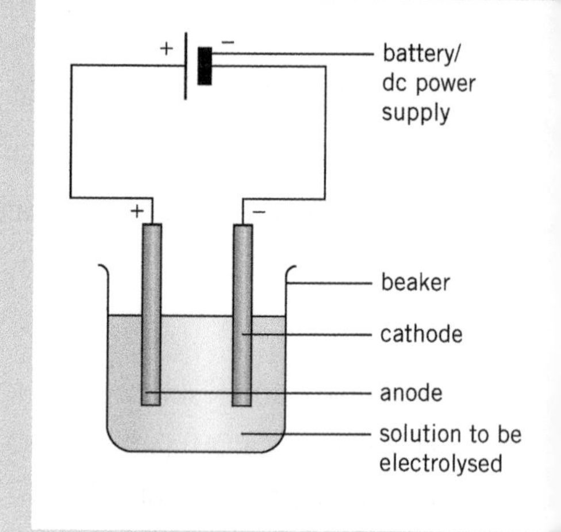

Remember !

This practical tests your ability to identify elements and compounds from observations. Remember that electrolysis uses electricity to break ionic compounds down into elements or simpler compounds. Metals or hydrogen are made at the negative electrode and non-metal molecules, including oxygen, are made at the positive electrode.

Exam Tip

There are many solutions that can undergo electrolysis. You need to be able to apply the principles of the practical you carried out to any example, including molten ionic compounds.

You should be able to predict the products of the electrolysis of different solutions and identify which ion goes to each electrode.

For higher tier, you should be able to write an equation for the reaction at each electrode.

1 Give the meaning of electrolysis. [2 marks]

____

____

2 Electrolysis involves the movement of ions towards electrodes.

Describe the difference between atoms and ions. [2 marks]

____

____

____

3 Draw one line from each statement to the electrode it describes. [2 marks]

Statement	Electrode
Positive electrode	
Negative electrode	Anode
Positively charged ions move towards it	Cathode
Negatively charged ions move towards it	

4 a Explain why electrolysis would not occur if the electrodes were touching each other. [2 marks]

____

____

____

b A student checks that the electrodes are not touching but still cannot see gas being produced at the electrodes.

Suggest an alteration to the experiment that would immediately show if the circuit was working properly. [1 mark]

____

____

5 Describe the hazards and/or risks associated with electrolysis of copper(II) chloride and the safety precautions that should be taken to reduce these risks. [6 marks]

6 Two students carried out electrolysis of copper(II) chloride.

- Student A set their power supply to 4 V.
- Student B used a power supply set at 1 V.

Their methods were the same in all other ways.

Compare the results you would expect from the two experiments. [4 marks]

Exam Tip

When 'compare' is the command word, you need to include things that are the same and things that are different.

7 In the two experiments described in the method, the copper ions and the sodium ions will both move to the same electrode.

Identify which electrode the copper ions and sodium ions will move towards and explain why. [2 marks]

Electrode = ______________________

Explanation = ______________________

8 The equipment in the diagram below is used to collect the gases produced during the electrolysis of calcium nitrate. [4 marks]

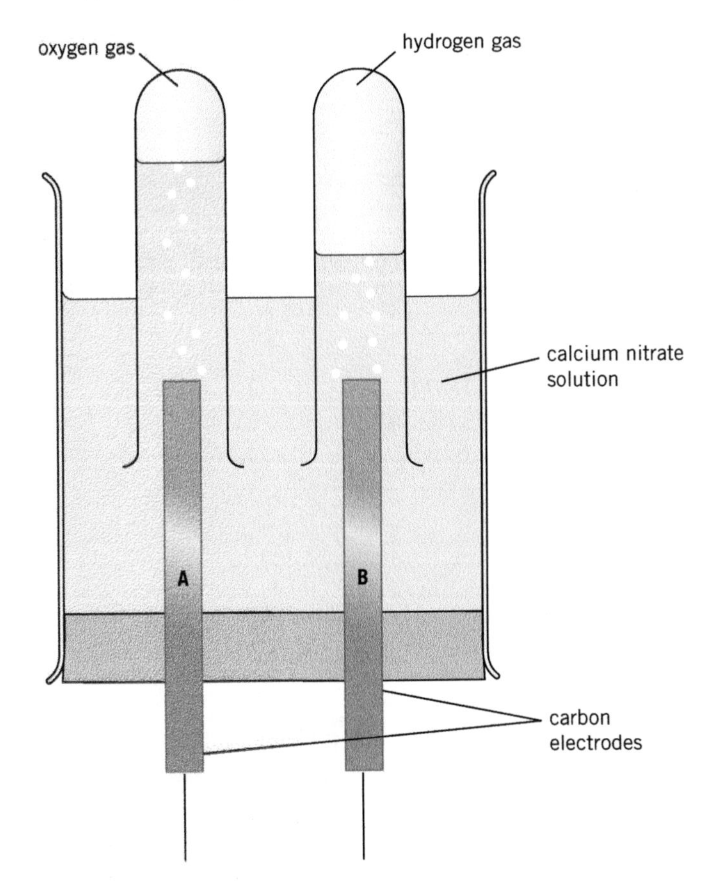

a Describe the tests you could carry out to confirm the identity of the two gases. [3 marks]

____

____

____

____

____

____

b Give the letter of the anode and the cathode. [2 marks]

Anode = __________

Cathode = __________

Hint

Think about the charges on each of the ions.

9 **a** In the electrolysis of sodium chloride solution, hydrogen gas is formed at one of the electrodes.

Give the reason why hydrogen gas is formed instead of sodium metal.

____

____

b There are three products from the electrolysis of sodium chloride; two gases and a compound in solution.

Give the formula of the compound in the solution. [1 mark]

Formula = ____________________

c Explain why solid sodium chloride cannot undergo electrolysis. [2 marks]

____

____

____

____

d Explain how and why you would expect the pH of the solution to change during sodium chloride electrolysis. [3 marks]

____

____

____

____

10 **H** Use ionic half equations to describe what happens at each electrode in copper(II) chloride electrolysis. [6 marks]

____

____

____

____

____

____

____

____

____

____

Hint

Look at the number of marks available. Six marks doesn't mean you need to write an essay, but you do need to write three things about each electrode.

11 Ⓗ A teacher demonstrates the electrolysis of lead bromide. Their experiment is shown in the diagram below.

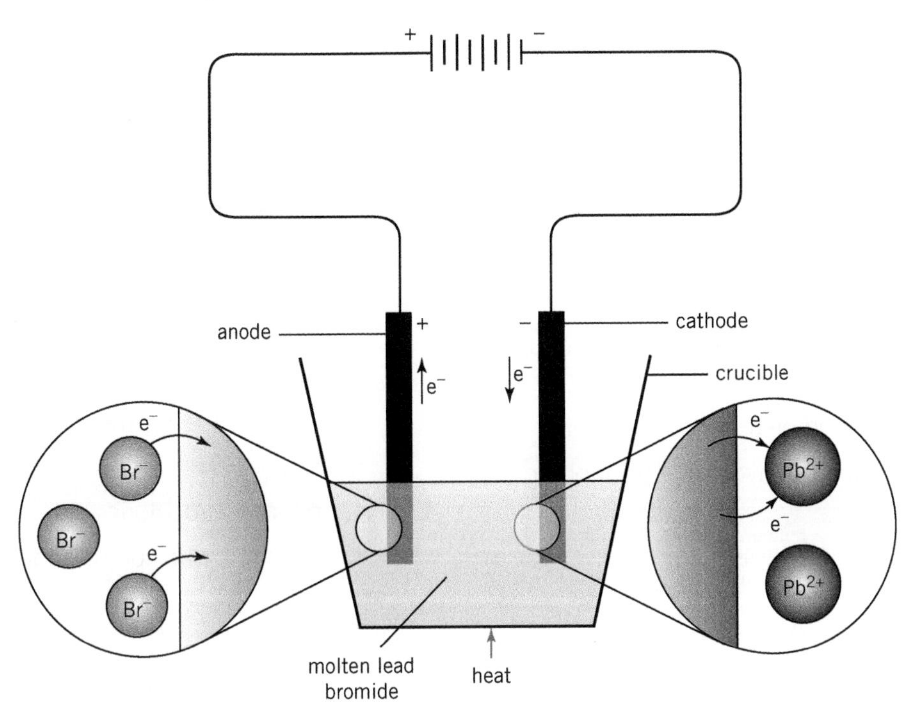

a Explain what will happen at each electrode in terms of oxidation and reduction. [4 marks]

____

____

____

____

____

____

____

____

____

____

b Suggest a safety precaution that should be taken while the teacher carries out the practical. [1 mark]

____

____

10 Temperature changes

Investigate variables that affect temperature change in reacting solutions

Method

1. Place the polystyrene cup inside a 250 cm³ beaker.
2. Measure 30 cm³ of dilute hydrochloric acid in the measuring cylinder and pour into the polystyrene cup.
3. Place the lid on the cup and insert the thermometer.
4. Record the temperature of the dilute hydrochloric acid.
5. Measure 5 cm³ of sodium hydroxide solution, pour it into the cup, and stir to mix.
6. Record the highest temperature reached by the thermometer.
7. Repeat steps 5 and 6, adding 5 cm³ sodium hydroxide solution up to a maximum of 40 cm³.
8. Repeat the entire experiment another two times.

Safety

- Sodium hydroxide solution: CORROSIVE
- Hydrochloric acid: IRRITANT
- Wear chemical splash-proof eye protection and wash hands after the practical.

Equipment

- eye protection
- 50 cm³ measuring cylinders and a 250 cm³ beaker
- polystyrene cup and lid with a hole for a thermometer.
- weighing boat, spatula, and balance
- 0–110°C thermometer
- stopwatch
- dilute hydrochloric acid
- sodium hydroxide solution

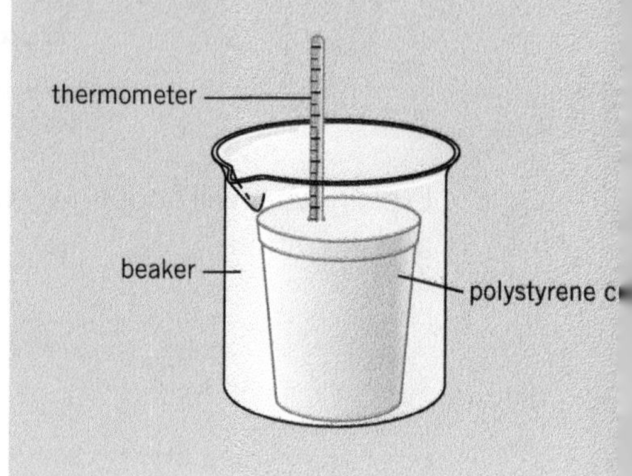

Remember

This practical tests your ability to safely and accurately measure mass, temperature, and volume in order to investigate chemical reactions. In this reaction you are mixing a strong acid with a strong alkali, as in Required Practical 8 Making salts. A key skill being tested is your ability to extract information from graphs.

Exam Tip

You will need to know:

- the general equations for the reactions of acids and be able to apply them
- how to determine the formula of ionic compounds from the charges on their ions
- formulae of all the ions involved in the neutralisation reaction.

1 a Give a reason for using a cup made from polystyrene. [1 mark]

____

____

b Give the function of the beaker. [1 mark]

____

____

c Give the function of the lid. [1 mark]

____

____

2 Give two possible sources of error in this experiment. [2 marks]

____

____

3 Complete the balanced equation for the reaction of sodium hydroxide solution (NaOH) with dilute hydrochloric acid (HCl). [2 marks]

________ (aq) + HCl(aq) ⟶ ________ (aq) + ________ (l)

Hint

Pay close attention to ALL the information being given to you in the equation.

4 Explain why it is important to wait until the reading on the thermometer stops changing before recording the temperature. [2 marks]

____

____

____

____

5 It is important not to leave long gaps between adding each 5 cm^3 sample of sodium hydroxide solution in this experiment.

Explain what effect adding the sodium hydroxide solution after a very long interval would have on the results. [2 marks]

____

____

____

6 The equipment must be washed thoroughly before repeating the whole experiment.

Why it is important to repeat the whole experiment?

Tick **two** boxes. [2 marks]

Calculate a mean ☐

Improve accuracy ☐

Reduce effect of random errors ☐

Spot anomalies ☐

Test reproducibility ☐

Exam Tip

Always double check how many boxes you are being asked to tick.

7 Explain briefly why it is important to wash out the equipment before you repeat the experiment. [2 marks]

____

____

8 **a** The reaction between sodium hydroxide and hydrochloric acid is exothermic.

Sketch a reaction profile for this reaction on the axes below. [2 marks]

energy

progress of reaction

Exam Tip

Remember that if you are asked to sketch a graph, you don't need to plot any points. Just show the correct overall shape of the graph.

b Describe what happens to the energy transferred in this reaction. [1 mark]

____

____

c Give another example of an exothermic reaction. [1 mark]

____

9 A student carried out an experiment to measure the energy change calculated from the results was +41.7 kJ/mol.

Describe what this tells us about the energy change in the reaction. [1 mark]

____

____

10 a Plot the given data below on the axes provided. [3 marks]

Volume of NaOH added in cm^3	Mean highest temperature reached in °C
0	19
5	22
10	25
15	27
20	29
25	31
30	34
35	33
40	30

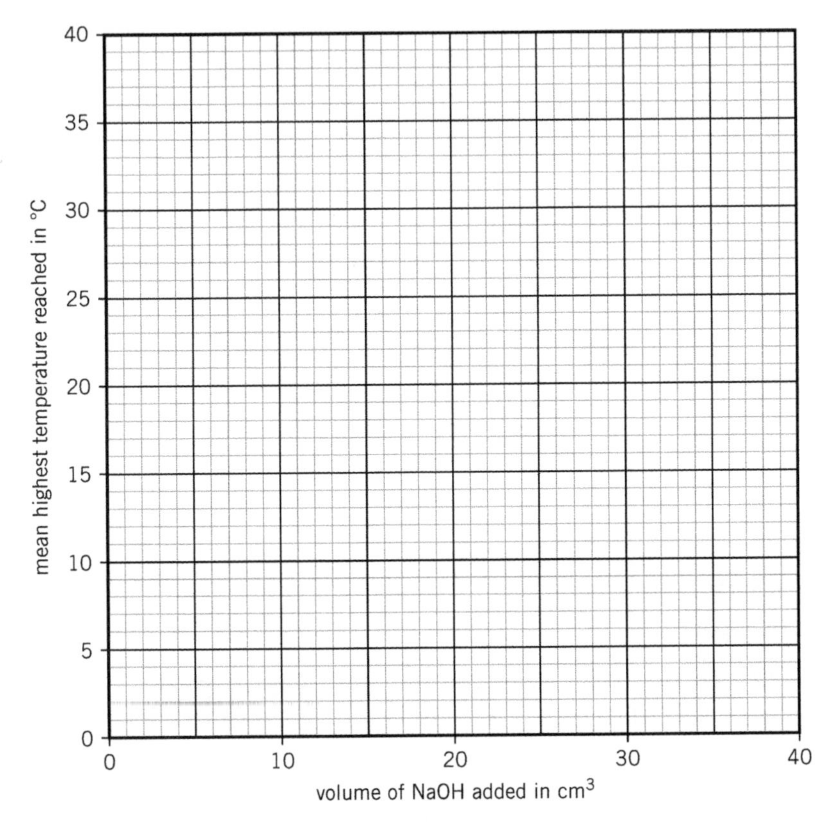

Hint

You will need to draw **two** lines of best fit on your graph – one as the temperature rises and the other as it falls.

b Estimate a value for the highest temperature reached. [2 marks]

Estimated highest temperature = ____________°C

c Describe the shape of the graph. [2 marks]

____

____

____

____

d Explain the shape of the graph. [3 marks]

____

____

____

____

____

____

Exam Tip

Describe and explain are very different command words. These two questions may look the same, but they have different answers. Describe is what it looks like, explain is why it looks like that.

e Suggest a piece of equipment that would allow the maximum temperature to be recorded without having to draw lines of best fit on a graph. [1 mark]

____

____

11 A student added sodium hydroxide solution in 10 cm^3 intervals and used their results to estimate the highest temperature change if 23 cm^3 of sodium hydroxide solution was added.

Describe how they could change the method to confirm their estimate.

[3 marks]

____

____

____

____

12 A student is carrying out two reactions:

- copper(II) sulfate solution and iron filings
- dilute nitric acid with potassium hydroxide.

a Describe a method the student could use to investigate whether the reactions are endothermic or exothermic. Include details of what the student would expect to see for each result. [6 marks]

Exam Tip

When being asked to suggest a method for an investigation, read what the aim is carefully. In this case the student is not necessarily interested in finding the maximum temperature rise.

____

____

____

____

____

____

____

____

____

____

____

____

In a similar experiment, 28 g of iron filings react with the copper(II) sulfate solution according to the equation:

$$Fe(s) + CuSO\ (aq) \longrightarrow FeSO\ (aq) + Cu(s)$$

b Calculate the relative formula masses of copper(II) sulfate and iron sulfate. [2 marks]

____

____

____

Relative formula mass of copper(II) sulfate = __________

Relative formula mass of iron(II) sulfate = __________

c (H) Calculate the number of moles of iron filings that react. [2 marks]

____

____

Number of moles of iron reacting = __________ moles

d (H) Calculate the mass of iron(II) sulfate produced in this experiment. [2 marks]

Mass of iron(II) sulfate produced = __________ g

e (H) Explain why some reactions are exothermic and some are endothermic. [4 marks]

Hint

This question expects you to talk about covalent bonds being broken and made.

____

____

____

____

____

____

13 (H) The reaction between sodium hydroxide solution and dilute hydrochloric acid is an example of neutralisation.

Write the ionic equation for this neutralisation reaction, including state symbols. [3 marks]

____

11 Rates of reaction

Investigate how changes in concentration of a reactant affect the rate of a reaction.

Method

A Measuring the production of a gas

1. Fill the water trough and the measuring cylinder with water, and clamp the cylinder upside down in the water trough.
2. Set up the conical flask, bung, and delivery tube so that the exit of the delivery tube is under the measuring cylinder.
3. Add 50 cm^3 of 2 mol/dm^3 hydrochloric acid into the conical flask.
4. Sandpaper 3 cm of magnesium ribbon, drop it into the conical flask, quickly replace the bung, and start the stopwatch.
5. Record the volume of gas produced every 10 seconds until no more gas is being produced.
6. Repeat steps 1 to 5, for each concentration of hydrochloric acid.

B Measuring reaction rate by change in turbidity

1. Add sodium thiosulfate solution and distilled water to the conical flask in the following proportions to make each concentration:

Volume of distilled water in cm^3	40	30	20	10	0
Volume of sodium thiosulfate in cm^3	10	20	30	40	50
Final sodium thiosulfate concentration in g/dm^3	8	16	24	32	40

2. Add 10 cm^3 dilute hydrochloric acid, place the conical flask on the black cross, and start the stopwatch.
3. Record the time when the black cross is no longer visible.
4. Repeat steps 1–3 for the other concentrations of sodium thiosulfate.

Safety

- dilute hydrochloric acid: IRRITANT
- Eye protection should be worn.

Equipment

- conical flasks
- rubber bung and delivery tube to fit conical flask
- water trough
- clamp stand, boss, and clamp
- 100 cm^3 measuring cylinders
- stopwatch
- dilute hydrochloric acid at different concentrations (between 0.25 and 2.0 mol/dm^3)
- 3 cm strips of magnesium ribbon
- sodium thiosulfate solution (40 g/dm^3)
- distilled water
- printed black cross
- sandpaper

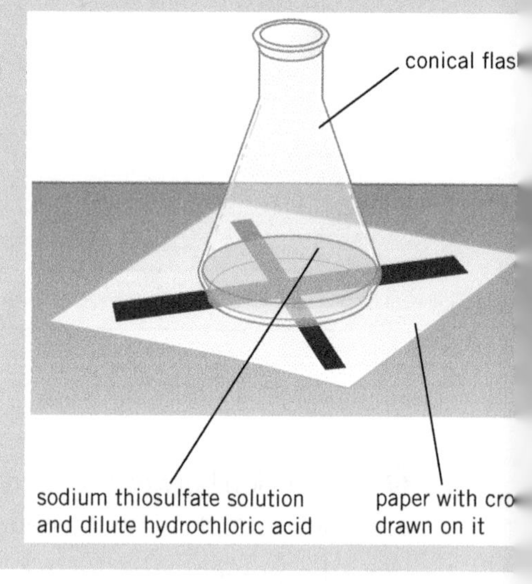

Remember !

This practical demonstrates two of the ways that the rate of a reaction can be measured. Remember that every method of measuring the rate of a reaction is actually measuring the decrease of a reactant, or the increase of a product. This could mean, for example, measuring the volume of gas forming, measuring mass lost as a solid turns to gas and escapes, or observing the formation of a coloured product.

Exam Tip 

There are many different methods that different schools will use for this practical. The important thing to remember is that the same principles are applicable across all of them.

The method for collecting gas with an inverted measuring cylinder is similar to the method for measuring the rate of photosynthesis in biology so similar questions could apply to both practicals.

This is a common practical question in exams – learn it well!

1 Give **three** ways of measuring the rate of a reaction. [3 marks]

____

____

____

2 Identify which of the following equations is correct.

Tick **one** box [1 mark]

$$\text{Mean rate of reaction} = \frac{\text{quantity of product formed}}{\text{time taken}}$$ ☐

$$\text{Mean rate of reaction} = \frac{\text{time taken}}{\text{quantity of product formed}}$$ ☐

mean rate of reaction = quantity of product formed × time taken ☐

mean rate of reaction = quantity of reactant used × time taken ☐

3 In an experiment, 56 dm^3 of carbon dioxide was produced in 15 seconds.

Calculate the mean rate of reaction.

Give your answer to 2 significant figures. [2 marks]

____

____

____

Mean rate of reaction = ______________ dm^3/s

4 This practical investigates the effect of reactant concentration on reaction rate.

Give **three** other factors that can affect the rate of a reaction. [3 marks]

5 Explain why it is important to rub magnesium ribbon with sandpaper before using it. [3 marks]

6 During the reaction between dilute hydrochloric acid and magnesium ribbon, hydrogen gas is released.

Describe the test for hydrogen gas. [1 mark]

7 A student carried out the experiment with dilute hydrochloric acid and sodium thiosulfate solution. After the first three concentrations they lost the printed cross the teacher had provided. They decided to draw their own replacement cross and carried on, starting with the fourth concentration.

Evaluate the student's decision. [3 marks]

8 The reaction between hydrochloric acid and sodium thiosulfate can be described by the following reaction.

$$\_\_HCl(aq) + Na_2S_2O_3(aq) \rightarrow \_\_NaCl(aq) + S(s) + SO_2(aq) + H_2O(l)$$

a Complete the equation above by balancing it. [1 mark]

Hint

This equation may look much more complicated than ones you have done before but notice that there are gaps showing where you need to add in the numbers. Only put numbers in these places.

b The products all have different state symbols.

Complete the table below with:

- the name of each product
- the state symbols
- the state of the product
- any observations that you might see as the product is formed.

[5 marks]

Product	State symbol	State of product	Observation as product is made
water	(l)	liquid	no change observed
sulfur	(____)	________	______________ ______________
________	(aq)	________	no change observed
________	(____)	aqueous	______________ ______________

9 A pair of students carried out these two methods in class. They took turns doing the different jobs each time they repeated the experiment.

Explain why it is important for the same person to do the following jobs in each repeat.

a Deciding when the cross has disappeared. [2 marks]

____

____

____

b Adding the magnesium ribbon to the hydrochloric acid and putting the bung on. [2 marks]

____

____

____

10 Give the dependent variable in each of the two experiments described in the methods. [2 marks]

Dependent variable in experiment 1 = ________________________

Dependent variable in experiment 2 = ________________________

11 A student reacts marble chips with hydrochloric acid and collects the gas produced in a syringe. They record the volume of gas in the syringe every 5 seconds. They repeat the experiment three times using different concentrations of hydrochloric acid:

- 0.5 mol/dm^3 HCl
- 1.0 mol/dm^3 HCl
- 2.0 mol/dm^3 HCl

Predict how the results would vary for the three concentrations and explain your prediction using collision theory. [6 marks]

12 A student carried out a reaction where she kept concentration the same but changed the temperature and drew the following a graph of her results.

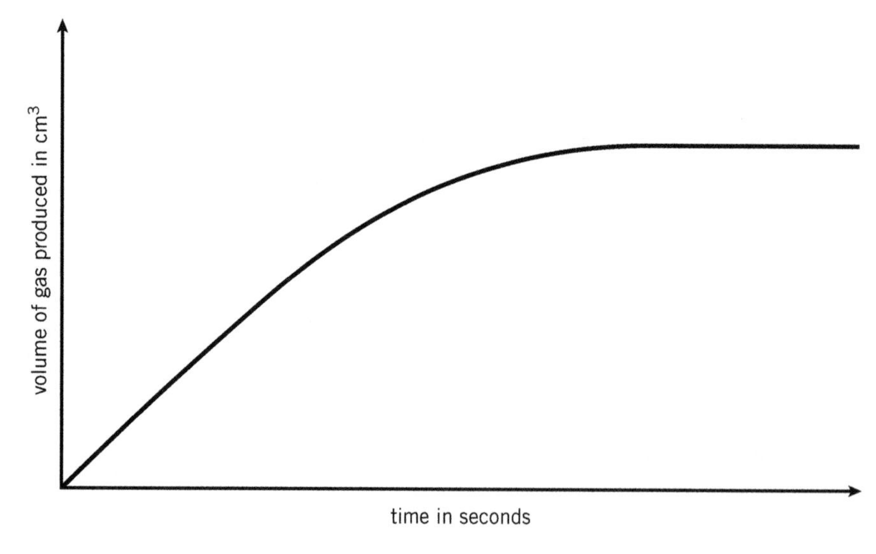

The student then repeated the experiment at a higher temperature.

a Sketch a line on the graph to show the results you would predict at a higher temperature. [2 marks]

b Explain your prediction. [3 marks]

13 A group of students carried out method **B** but could not agree exactly when the cross they were watching disappeared.

Suggest an alternative method that would reduce the amount of error in the results. [1 mark]

14 Explain why it is important to decrease the volume of water added when increasing the volume of sodium thiosulfate. [1 mark]

15 20 g sodium thiosulfate is dissolved in 0.5 dm^3 distilled water.

What is the concentration of the solution?

Tick **one** box. [1 mark]

0.5 mol/dm^3 ☐

10 g/dm^3 ☐

20 g/dm^3 ☐

40 g/dm^3 ☐

16 Marble chips reacted with hydrochloric acid in a flask. The flask was placed on a scale and the mass recorded over time.

a Plot the following results on the axes provided below. [4 marks]

Time in s	Mass of flask in g
0	95
10	43
20	27
30	21
40	20

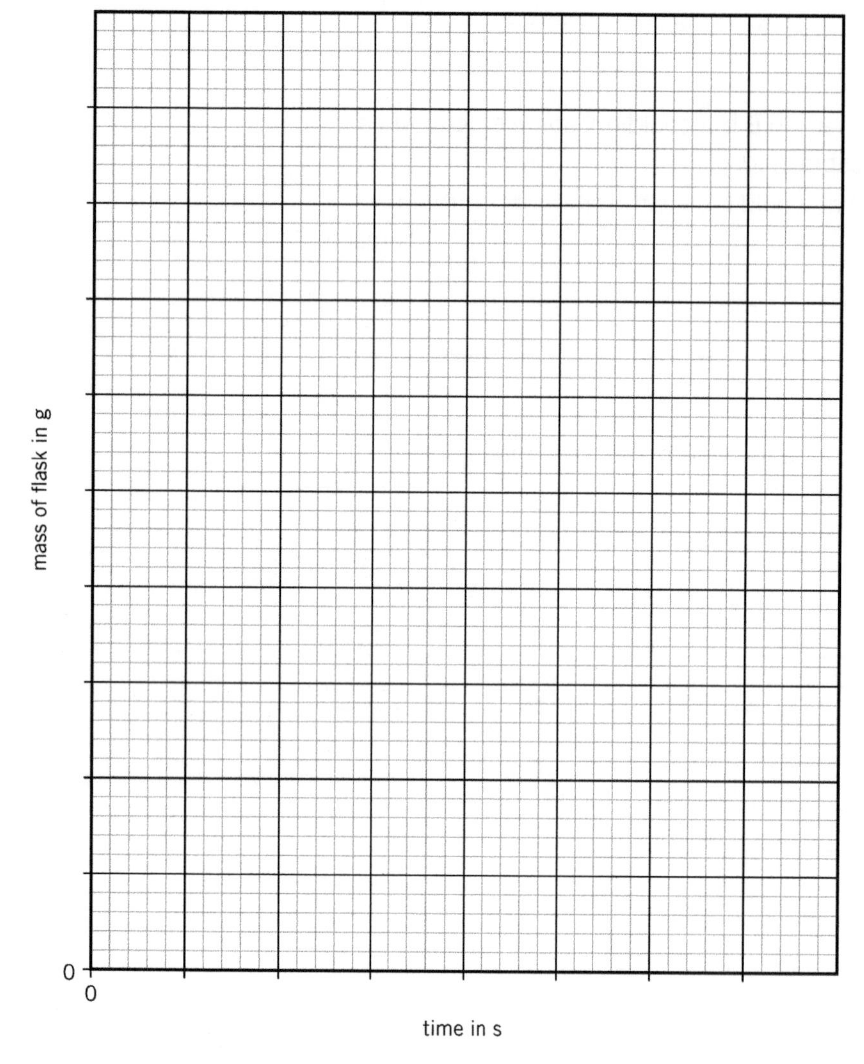

b Draw a line of best fit on the graph. [1 mark]

c Compare the rate of reaction at 10 seconds and at 30 seconds. [2 marks]

____

____

____

Exam Tip

If you are asked about the rate at a point on a curved graph, you need to draw a tangent to the curve at that point. Remember, the steeper the tangent, the faster the rate.

d **i** Write the equation to calculate the mean rate of reaction in g/s.

[1 mark]

ii Calculate the mean rate of reaction in g/s between 0 s and 40 s.

[2 marks]

____

____

Mean rate of reaction between 0 s and 40 s = ____________ g/s

e **H** Calculate the rate of reaction at 20 s:

i in g/s

[2 marks]

____

____

____

Rate of reaction at 20 s = ____________ g/s

ii in mol/s

[2 marks]

____

____

____

____

Rate of reaction at 20 s = ____________ mol/s

Hint

You are simply being asked to convert g/s into mol/s. This is basically the same as converting g of a substance to mol of a substance.

Think about what substance is being lost from the container. What is its relative formula mass?

12 Chromatography

Use chromatography to separate and tell the difference between coloured substances.

Method

1. Use a pencil to draw a horizontal base line, 1 cm from the bottom of the chromatography paper.
2. Use a pencil to draw a cross on the centre of the base line.
3. Use a thin paint brush or capillary tube to add some of the food colouring onto the cross and allow it to dry.
4. Fold the top edge of the chromatography paper over a wooden splint and keep in place with a paper clip.
5. Add 0.5 cm depth of water into the beaker.
6. Carefully lower the chromatography paper into the beaker, taking care to keep the pencil line above the water level. Leave until the water line (solvent front) has passed the last coloured spot.
7. Remove the chromatogram and allow it to dry.

Equipment

- food colourings
- capillary tubes
- chromatography paper
- pencil and ruler
- water
- 250 cm^3 beaker
- paper clip and wooden splint

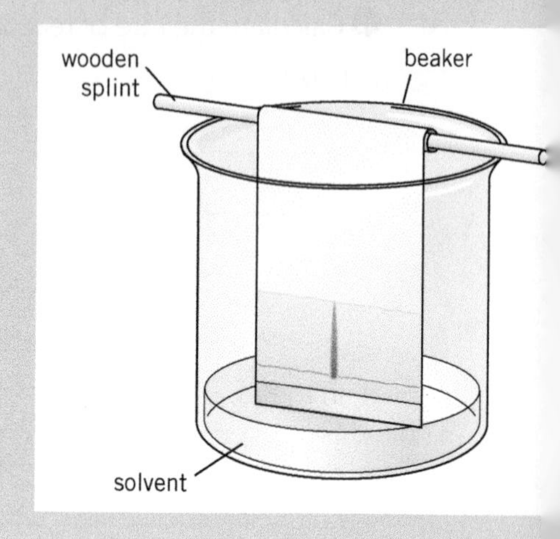

Safety

- Do not eat the food colouring.
- Make sure you are aware of anyone with food colouring allergies.
- Capillary tubes can be fragile and care should be taken so they do not break.

Remember !

This practical tests your ability to separate coloured substances by making paper chromatograms. Food colourings are often made by mixing different coloured substances. The mixture of substances can be separated using chromatography. You can identify the separated substances by comparing how far they move compared to samples of known inks and dyes.

Exam Tip

Make sure you know how to calculate R_f values

$$R_f \text{ value} = \frac{\text{distance moved by solute}}{\text{distance moved by solvent}}$$

Remember that the solute is the spot and the liquid in the beaker is the solvent. The distance moved by solvent is the solvent front.

Chromatography involves a stationary phase and a mobile phase. In this experiment the paper is the stationary phase and the water is the mobile phase.

1 Give the meaning of the term 'mixture'. [1 mark]

____

____

2 Chromatography is one way that we can determine whether a substance is pure or not. Give two other methods can we use to determine the purity of a substance. [2 marks]

Method 1: ______________________________

Method 2: ______________________________

3 a Label the apparatus shown below with the terms in the box. [4 marks]

solvent	solvent front	R_f value = 0.34	R_f value = 0.60

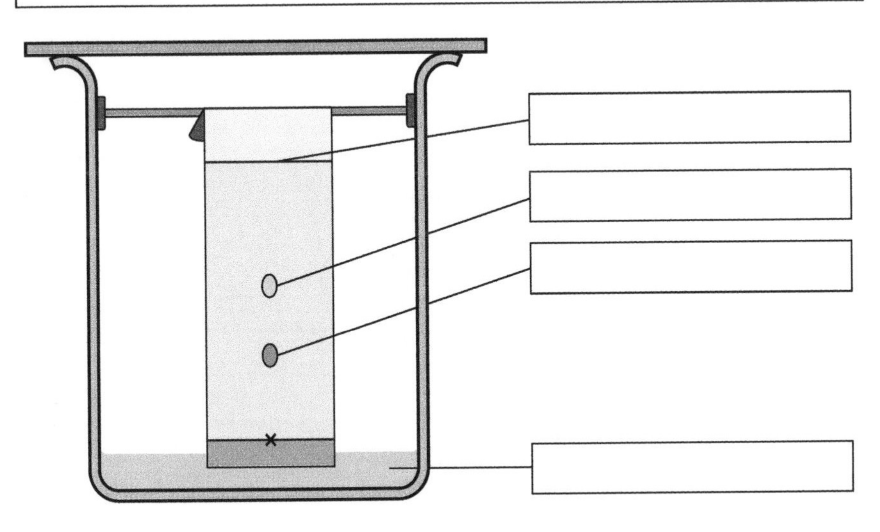

b Suggest the main purpose of the lid in the diagram in part **a**.

Tick **one** box. [1 mark]

Keeps impurities out of solvent ☐

Makes the solvent run faster ☐

Makes the solvent run slower ☐

Stops the solvent evaporating ☐

4 Chromatography can be used to test whether certain known substances are present in a sample.

Five different food colouring samples (A–E) are compared to red, blue, and yellow reference samples. The results are shown below.

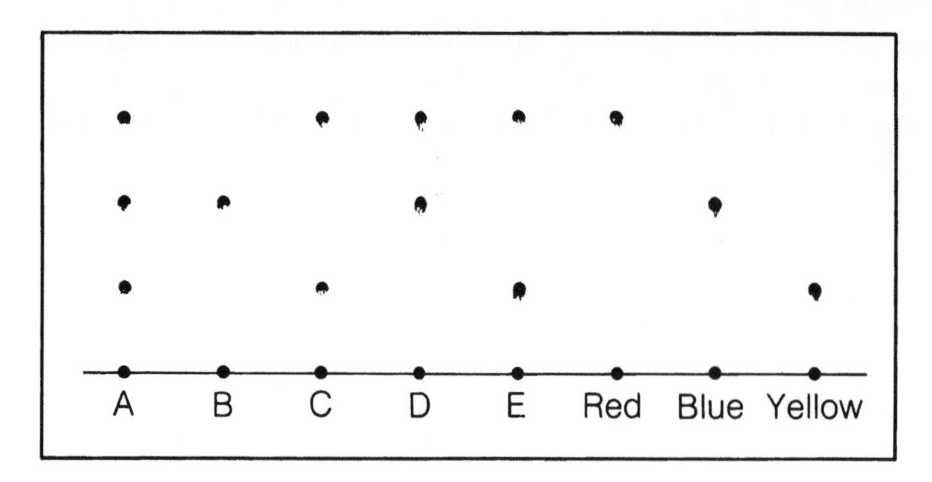

a Use evidence from the diagram to suggest which colour sample B is. [1 mark]

Colour of sample B = ______________

b How many different substances make up sample A? [1 mark]

Number of substances in sample A = ______________

c Give the letters of two unknown samples that are actually the same mixture. [2 marks]

Samples ______________ and ______________

d Use a ruler to draw a line on the diagram showing where the solvent should be in relation to the origin line at the start of the experiment. [1 mark]

e Explain why it is important to draw the origin line in pencil. [2 marks]

5 a Using the diagram below, determine which two athletes have been using banned substances. [2 marks]

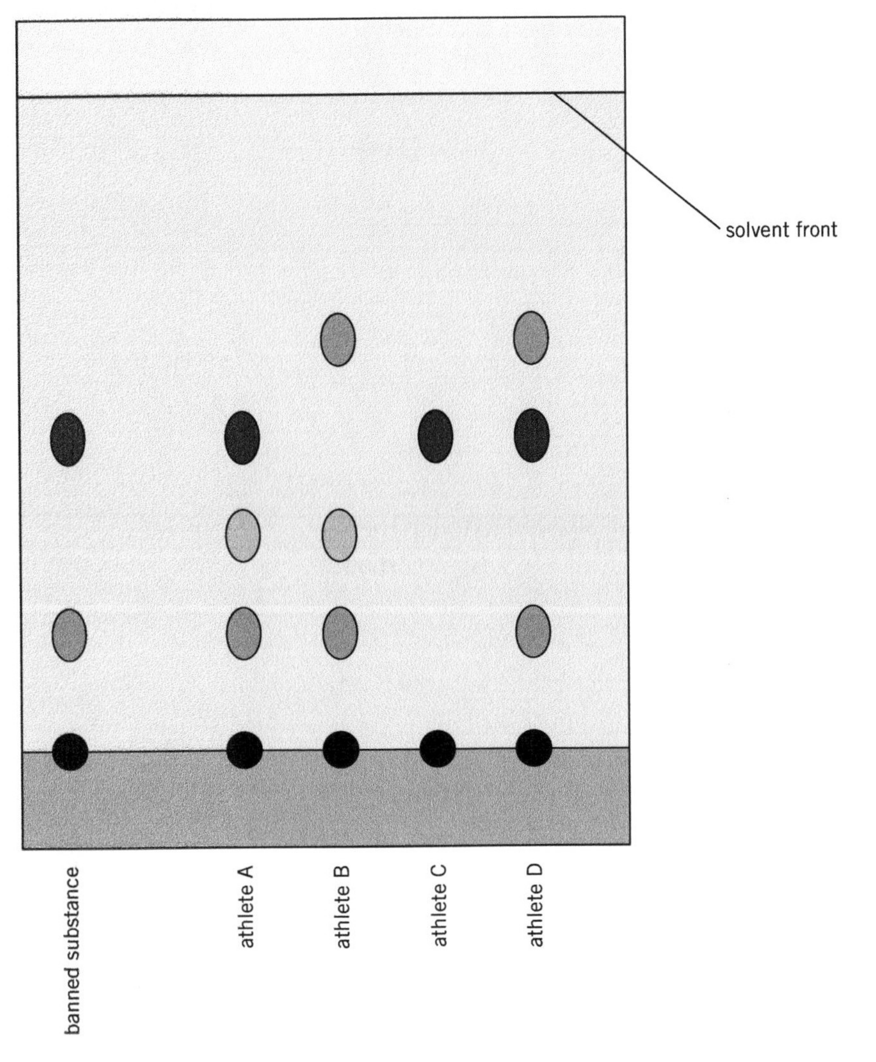

Athletes using banned substance = ______________ and ______________

b Give the equation used to calculate R_f values. [1 mark]

____

____

____

c Calculate the R_f value of the spot in sample C. [2 marks]

____

____

____

R_f value of spot in sample C = ______________

Hint

Make sure you measure from the middle of the spot.

6 Two groups of students carried out this practical.

Group A allowed solvent to move up until it had reached 75% of the way up the paper. Group B were worried about the spots spreading out too much so stopped the experiment when the solvent front was only 25% of the way up the paper.

Evaluate the methods used by the two groups. [3 marks]

7 Explain why the chromatography paper should not be allowed to touch the sides of the beaker. [2 marks]

8 Compare what you would expect to observe when a soluble sample and when an insoluble sample are spotted on the chromatography paper.

[3 marks]

9 Large chemical companies can use chromatography to test their products for purity. Suggest why a company might send samples to be tested at an independent lab. [3 marks]

10 Databases of R_f values also contain information about the solvent, and the temperature used to generate the R_f values.

Explain why it is important to use the same solvent and temperature as given in the database if you want to positively identify unknown substances from a chromatogram. [4 marks]

13 Water purification

Analyse and purify water samples.

Method

A Analysis

1. Use universal indicator paper to test the pH of the water sample.
2. Weigh an evaporating basin, recording the mass to 2 dp.
3. Add 5 cm^3 of the water sample to the evaporating basin and heat over a tripod and gauze until the water has evaporated.
4. Allow the evaporating basin to cool before weighing it again.
5. Subtract the original mass of the dish from the new mass to calculate the mass of dissolved, solid impurities in the water sample.

B Purification

1. Set up the apparatus as shown in the diagram in the equipment section.
2. Adjust the height of the thermometer so that the bulb is in line with the opening of the delivery tube.
3. Ignite the Bunsen burner with the air hole closed.
4. Open the air hole of the Bunsen burner so the flame turns blue and move the Bunsen burner under the tripod to heat the solution.
5. Note the temperature on the thermometer when it is at a constant value. This is the boiling point of the distillate.
6. Once half a boiling tube of distillate has been collected, remove the delivery tube and turn off the Bunsen burner.
7. Use universal indicator paper to test the pH of the distillate.

Equipment

- water sample
- Bunsen burner
- flame proof mat, tripod, and gauze
- conical flask
- two-hole bung
- universal indicator paper or pH probe
- spirit thermometer −10–110 °C
- delivery tube
- boiling tube
- large beaker with crushed ice
- anti-bumping granules
- evaporating basin
- balance

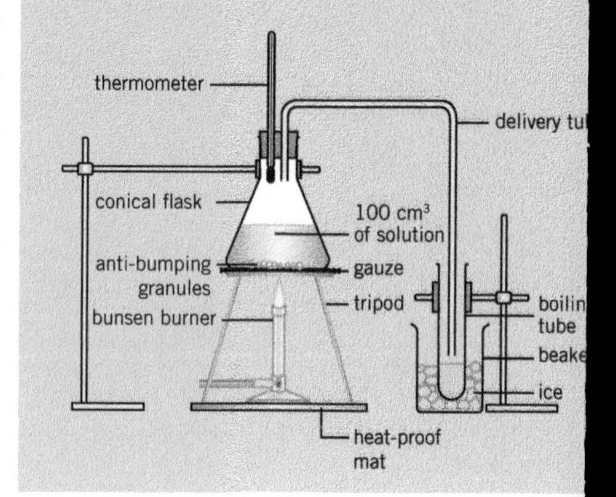

Safety

- Remember that the blue flame of the Bunsen burner is for heating and can cause burns.
- The glassware will get hot; make sure it cools before you touch it.
- Steam from the boiling mixture can cause scalds.
- Wear eye protection.

Remember !

The first part of this practical is about analysing a water sample to find out how pure it is. This is done by testing the pH and working out what the mass of any solids dissolved in the water is. The second part is purifying the sample using distillation. You need to be clear about which techniques are for analysing and which are for purifying and you should be able to describe them in detail.

Exam Tip 

The formula of water is H_2O. When you're writing this, the size of the number and letters is important. You will likely lose marks of you write H2O or h2o or H^2O.

1 a Circle the pH of pure water.

3	4	5	6	7	8	9	10	11

b Which of the following methods can be used to measure the pH of a solution?

Tick **two** boxes. [2 marks]

Blue cobalt chloride paper ☐

Boiling point ☐

Litmus paper ☐

pH probe ☐

Melting point ☐

Universal indicator solution ☐

2 Give definitions of the terms 'pure water' and 'potable water'. [2 marks]

Pure water ______________________________

____

____

Potable water ______________________________

____

____

3 Complete the table using terms from the box below. [3 marks]

desalination	distillation	filtration	purification	sterilisation

Name of process	Application
______________	Kill or remove harmful microbes.
______________	Remove insoluble particulates from sample.
______________	Separate liquids with different boiling points.

4 The diagram below shows a student's distillation experiment.

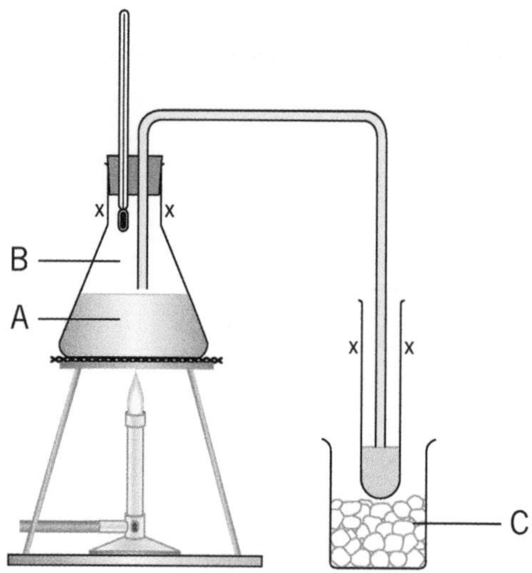

a Describe the differences in the arrangement and movement of water particles at points A, B, and C. [6 marks]

Exam Tip

Your answer doesn't always need to be all writing. A carefully annotated diagram can also gain marks in some situations.

b Use your knowledge of particle theory to explain what happens to the particles in the water sample as it is distilled. [4 marks]

Hint

This question appears to be very similar to part **a**. Make sure you really read the question and think about exactly what it is asking you. Part **a** only wanted you to describe the particles at each point. Part **b** wants you to explain the processes in terms of particle theory.

c Explain two things the student has done wrong when setting up the equipment shown in the diagram. [4 marks]

____

____

____

____

____

____

____

____

____

____

Hint

The command word is 'explain', so you should provide a reason why each thing the student has done wrong is important.

5 A student labels two solutions A and B but forgets to write down which one is which.

They know the two solutions are:

- 1 mol/dm^3 sodium chloride (NaCl)
- 1 mol/dm^3 sodium hydrogen carbonate ($NaHCO_3$)

After evaporating the water, sample A was found to have 0.585 g of salt in it and sample B had 0.84 g of salt in it.

Hint

Don't do more calculations than necessary. The solutions are the same concentration so you only need to know which solute has the greater relative formula mass.

a Identify which sample is 0.5 mol/dm^3 sodium bicarbonate and explain how you know. [3 marks]

____

____

____

____

____

____

b Suggest another way of identifying the two solutions that doesn't require the liquid to be evaporated. [1 mark]

____

____

____

6 The boiling point of water is 100 °C. The boiling point of ethanol is 78 °C. Both are colourless liquids.

Describe a method you could use to separate a mixture of 25% ethanol and 75% water. [6 marks]

____

____

____

____

____

____

____

____

____

____

7 A student investigates the purity of three unknown samples. Their results are shown in the table below.

	Boiling point in °C	pH	Mass of solid residue after evaporation of 100 ml solution in g	Conducts electricity?
distilled water	100.0	7.0	0.00	No
sample A	100.5	8.1	2.91	Yes
sample B	102.2	7.01	3.82	Yes
sample C	98.3	7.24	0.00	No

a Identify which sample has the most impurities dissolved in it and give a reason for your choice. [2 marks]

Sample with most dissolved impurities = ______

Reason __________

____

b The student says that measuring pH alone is not enough to measure the purity of a water sample.

Give evidence from the table that supports this statement. [2 marks]

____

____

____

c Sample C has no solid residue left when evaporated.

Suggest why it has a boiling point 1.7 °C lower than distilled water. [1 mark]

d The solid residue in samples A and B is found to be sodium chloride.

Explain why samples A and B conduct electricity, but distilled water does not. [2 marks]

8 Describe what happens to blue cobalt chloride paper when it is exposed to water. [1 mark]

Hint

Look at the number of marks available for each question. This is only worth one mark so it is only asking for a simple answer – not a whole paragraph.

14 Specific heat capacity

Determine the specific heat capacity of a material.

Method

1. Check the mass of the block of material you are using (it may be written on the block, or your teacher will be able to tell you).
2. The block should be wrapped securely in insulation and placed on an insulating mat.
3. With the power supply switched off, set up the apparatus as shown in the diagram.
4. Place the thermometer in the block and measure the temperature. Record this as the 'starting temperature' of the block.
5. Switch the power supply on and start the stopwatch.
6. Record the current and potential difference.
7. Watch the reading on the thermometer. Record the temperature every 60 seconds for ten minutes.
8. When the temperature reaches 15°C above the starting temperature, switch off the power supply and stop the stopwatch.
9. Record the thermometer reading and the time on the stopwatch.

Note: The thermometer reading might continue to increase for up to a few minutes after the heater has been switched off. Measure and record the highest reading of the thermometer after the heater was switched off.

Safety

- Do not touch electrical equipment, plugs, or sockets with wet hands.
- Do not touch the heater: it becomes very hot when in use, and can stay hot for a long time after it is switched off.
- Switch the heater off if you think it is overheating.
- Always switch the heater off when you are not using it.
- When the thermometer is not being used, make sure it is placed where it cannot easily roll off the table.

Equipment

- 12V, 24W low-voltage heater
- 12V power supply for heater
- ammeter and voltmeter
- two connecting leads
- aluminium block with holes for a heater and a thermometer
- insulation for block (including a thick insulating mat to place under it)
- rubber bands or tape to fix insulation around block

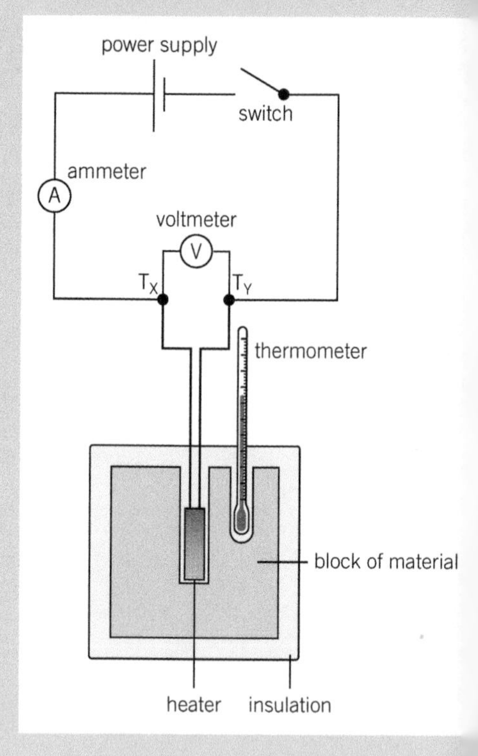

Remember !

This practical tests whether you can find out the specific heat capacity of a material by measuring current, potential difference, time and temperature. You should be able to describe the method, including details of how to use your results and any calculations you need to do.

Remember that a graph of temperature against work done should produce a straight line and the gradient of this line can be used to calculate the specific heat capacity.

Exam Tip 

There are several equations used in this practical and you need to be able to apply all of them. You will also need to recall the units for everything.

Many people were surprised a few years ago when the last question on a Physics paper was simply 'Give the units for specific heat capacity.'

power = current × potential difference **(need to be able to recall)**

energy = power × time **(need to be able to recall)**

energy transferred = mass × temperature change × specific heat capacity **(provided on data sheet)**

1 The equation below can be used to calculate the energy transferred in joules into the block of metal in a certain number of seconds.

Energy (J) = power x time (s)

Give the unit of power in this equation. [1 mark]

Unit of power = ______________

2 As part of this experiment you needed to set up a circuit including an ammeter and a voltmeter.

Tick the **two** correct statements that describe how they should be set up. [2 marks]

An ammeter should be connected in series ☐

An ammeter should be connected in parallel ☐

A voltmeter should be connected in series ☐

A voltmeter should be connected in parallel ☐

3 Two students investigated the specific heat capacity of identical blocks. Student A insulated their block with polystyrene. Student B did not use any insulation.

Explain which student's results would be more accurate. [4 marks]

____

____

____

____

____

____

____

____

Exam Tip

Just because there are eight lines provided to write your answer in, doesn't mean you have to fill them all.

It's much better to look carefully at the number of marks available. In this case, you should make four points.

4 A student investigated the specific heat capacity of bronze. They measured how many seconds it took for the temperature of the block to increase by 1°C.

Explain how the student could make their measurement more accurate.

[2 marks]

Exam Tip

Remember that the command word 'explain' means you need to give reasons in your answer.

5 A student uses a stopwatch that shows the time in minutes and seconds.

Convert 3 minutes 15 seconds into seconds. [1 mark]

Answer = ____________ s

6 Define specific heat capacity. [1 mark]

7 Specific heat capacity can be found using the following equation:

$$\text{specific heat capacity} = \frac{\text{energy transferred}}{\text{mass} \times \text{temperature change}}$$

Rearrange the equation so it could be used to find the temperature change.

[2 marks]

8 The above equation refers to energy transferred but when a graph is drawn for this experiment we use work done instead of energy transferred.

work done = force × distance.

Using this information, give 5 J in Nm. [1 mark]

Hint

Think about what the units for force and distance are.

9 Explain why the hole where the thermometer is placed is filled with water. [2 marks]

10 In this practical you used a thermometer and took a reading every minute to follow the temperature change. An alternative would be to use a temperature probe with a data-logger.

a Give two advantages of using a data-logger. [2 marks]

b Suggest a disadvantage of using a data-logger. [1 mark]

11 a 1 kg blocks of copper, steel, and aluminium are heated using the same amount of power.

Use the data in the table to explain which block's temperature will take the most time to increase by 30°C. [2 marks]

Metal	Specific heat capacity in J/kg °C
copper	385
steel	452
aluminium	913

b A student carried investigated the specific heat capacity of an unknown metal. They recorded the following data.

Time in s	Temperature in °C	Work done in kJ
0	21	0
60	22	1.8
180	28	5.4
300	36	9
420	44	12.6
540	52	16.2
600	56	18

Hint

The data for work done is given in kJ in the table. The standard unit for work done is J. You have to draw your graph in kJ but make sure that you use J when you do the calculation.

i On the axes provided below:

- plot these results
- draw a line of best fit. [3 marks]

temperature in °C: 0, 10, 20, 30, 40, 50, 60

work done in kJ: 0, 5, 10, 15, 20

ii Use the data table in part **a** and your graph to determine the identity of the metal. [1 mark]

____

____

c The student who recorded the data in part **b** has not recorded the data for work done correctly.

Describe how the data should have been recorded and explain why this change is important. [2 marks]

____

____

____

____

12 A student draws a graph of their results and notices that the first part of the graph is curved. This curve is followed by a straight section.

Suggest why the first part is curved. [2 marks]

____

____

____

____

Hint

The specific heat capacity can be found from the gradient of the **straight** section of the graph. You can then compare this with the data in **11a** to find the closest match.

Exam Tip

Always use the largest triangle you can draw to get an accurate gradient, and remember:

$$\text{gradient} = \frac{\text{change in up } (y)}{\text{change in across } (x)}$$

Hint

It's not the units.

13 (H) A student carried out the same experiment with a block of copper of unknown mass.

Use the data below to calculate the mass of the block.

Give you answer to an appropriate number of significant figures. [6 marks]

Current in circuit = 1.2 A

Potential difference in the circuit = 12 V

Temperature change at 4 minutes = 8°C

Answer = ______________ kg

Hint

You'll need to use all the equations in the 'key points'. To keep things simple, you'll need to use them in the order they are given.

Not all of the data is given in standard units so you'll need to do some converting before you start calculating.

Exam Tip

Sometimes you might need to look at earlier parts of a question to find the information you need. For example, the specific heat capacity for copper is given is **11a**.

15 Resistance

Investigate factors that affect resistance in electrical circuits.

Method

A The effect of the length of a wire on resistance

1. Set up the apparatus as shown in Figure 1.
2. Set the length of the test wire to 100 cm by adjusting the positions of the crocodile clips.
3. Turn the power supply on at 1.5 V and close the switch.
4. Record the readings of voltage and current.
5. Repeat steps 2 to 4, decreasing the length of the test wire by 10 cm each time to a minimum of 20 cm.

B Resistance in series and parallel circuits

1. Set up two resistors in series as shown in Figure 2a.
2. Close the switch and record readings of voltage and current for the series circuit.
3. Set up the two resistors in parallel as shown in Figure 2b.
4. Close the switch and record readings of voltage and current for the parallel circuit.

Equipment

- piece of wire to test
- low-voltage power supply or batteries
- switch
- connecting leads
- crocodile clips
- metre ruler
- ammeter and voltmeter
- wire-wound resistors

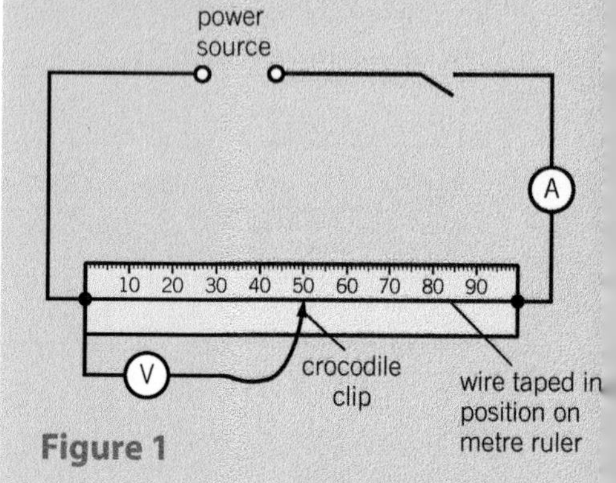

Figure 1

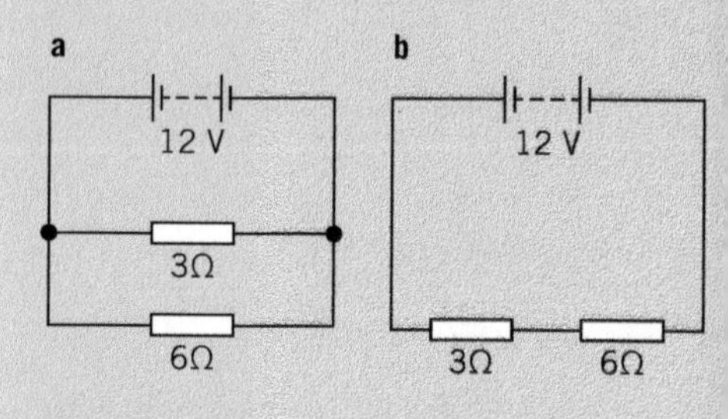

Figure 2

Safety

- Do not connect the wire directly to the mains supply – this could cause a fatal electric shock.
- Do not let the current get higher than 1.0 A. Use a variable resistor to keep the current low if necessary.
- Disconnect the circuit as soon as the measurements are taken, to stop the wire getting hot.
- After the circuit has been on, do not touch the wire without checking first whether you can feel heat coming off it from a distance, using the back of your hand.

Remember 

There are two separate experiments in this practical. One investigates how resistance across different lengths of wire varies. The other investigates how the arrangement of resistors in series and parallel circuits affects the resistance in the circuit. You should understand and be able to describe how to accurately measure current, potential difference, and resistance in both experiments.

As part of this practical, you are expected to be able plot and interpret a graph of resistance against length.

Exam Tip

For a series circuit the total resistance in a circuit is the sum of all the resistors. Make sure you learn the equation:

$R_{total} = R_1 + R_2 + R_3 \ldots$

You need to know and be able to apply the equation:

potential difference = current × resistance

1 Use the words in the box below to complete the following sentences. [2 marks]

ammeter	current	potential difference	voltmeter

In this experiment, the _______________ measures the _______________ flowing through the circuit.

The _______________ measures the _______________ across the resistor.

2 A student set up the following circuit.

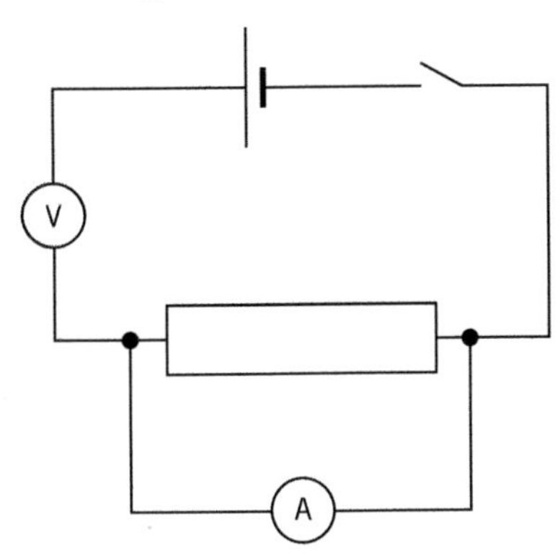

a Identify the error the student has made in setting up the equipment. [1 mark]

____

b When the student corrects the error, the ammeter and voltmeter both read zero.

Suggest what they need to change to be able to take readings. [1 mark]

____

c When the student gets the circuit working the ammeter reads 0.50 A and the voltmeter reads 0.75 V.

Calculate the resistance of the resistor and choose the correct units from the box below. [3 marks]

A	Ω	V	R	I

____

____

____

Resistance = ________ units = ______

3

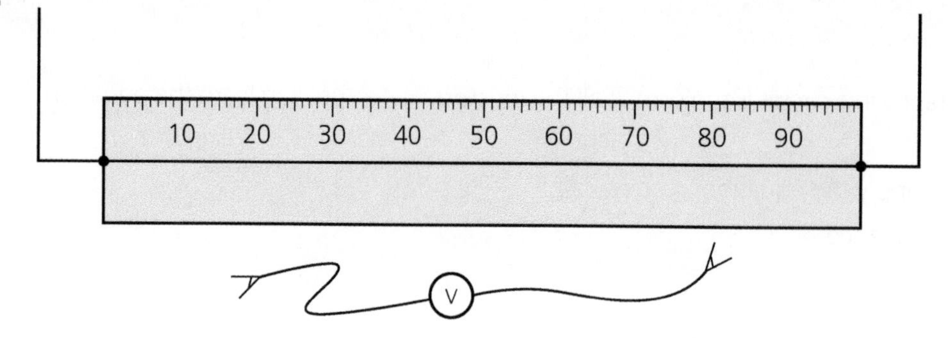

a A student wants to record the resistance across a 0.2 m length of wire.

On the diagram above, draw an X at each point where the student should clip the crocodile clips that are attached to the voltmeter. [1 mark]

Hint

Make sure you remember to convert the units.

b A 'zero error' is often seen in this experiment.

Describe a potential source of a zero error in this experiment. [2 marks]

____

____

____

c Sketch a graph on the axes provided to show the results you would predict in this experiment. [1 mark]

Exam Tip

When the command word is 'sketch', you only need to draw a line to show the overall shape of the graph. You don't need to plot any individual points.

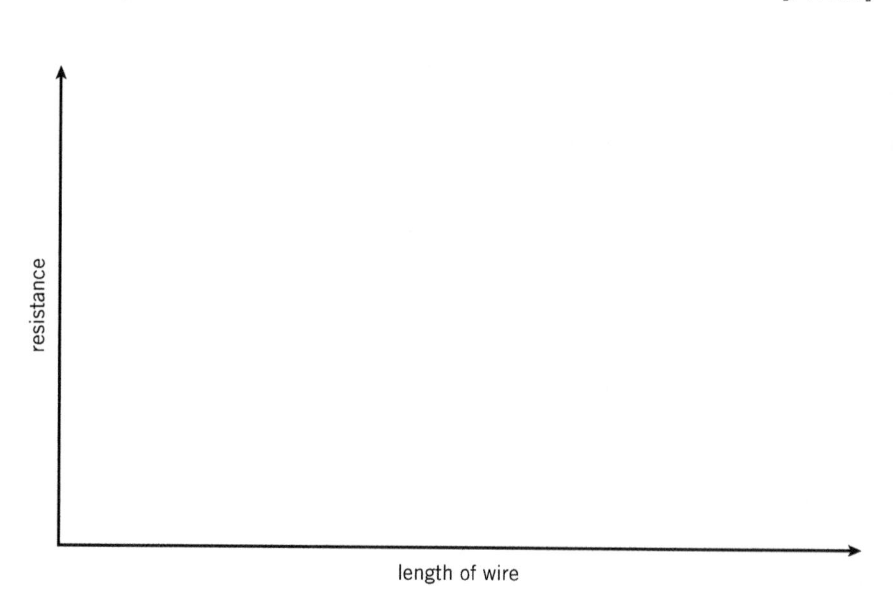

4 A student investigated the effect of the length of a wire on its resistance. The student repeated their experiment three times. Their results are shown below.

Length of wire in cm	Resistance in ohms			
	Test 1	Test 2	Test 3	Mean
20	4.02	4.10	4.62	
40	8.24	8.36	8.97	
60	12.60	15.67	12.98	
80	16.13	16.27	16.84	
100	19.04	19.25	19.99	

a Circle the anomalous result in the table. [1 mark]

b Calculate the mean for each length of wire to complete the table. [3 marks]

Exam Tip

Remember not to include any anomalous results when calculating a mean.

c The student did not allow the wire to cool down between each repeat experiment.

Explain the effect this had on the repeatability of the student's results. [3 marks]

Hint

You should quote examples from the results in your answer.

d • Use the data in the table to plot a graph of mean resistance in ohms against length of wire in cm.

• Draw a line of best fit. [4 marks]

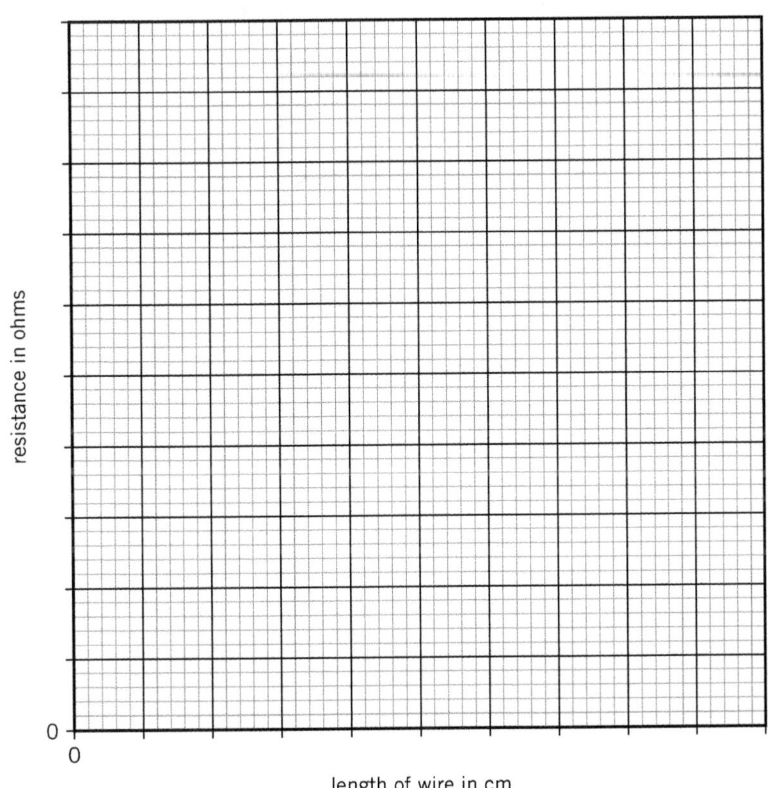

5 a In another experiment the length of the wire was kept constant and the cross-sectional area of the wire was changed.

Predict the effect that doubling the cross-sectional area of the wire would have on the resistance of the wire. [1 mark]

____

____

____

____

b A variable resistor is often used in this experiment to keep the current below 1.0 A.

Explain why the resistance of a wire will increase if the circuit is left on for a long time with a high current flowing. [4 marks]

____

____

____

____

____

____

Exam Tip

This is an 'explain' question, so it wants you to give reasons why something happens.

6 The diagrams below show the two components arranged in series and in parallel.

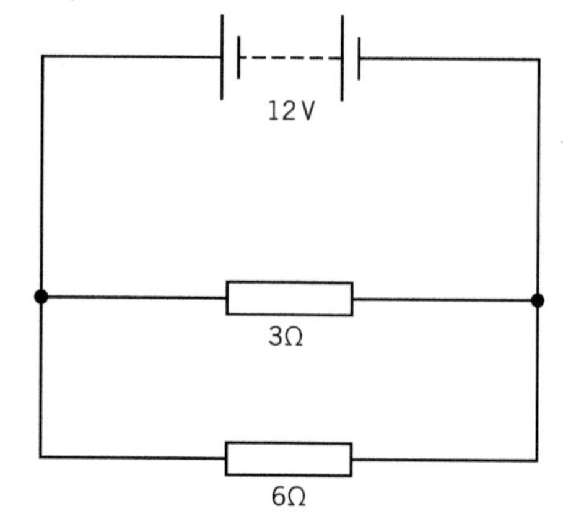

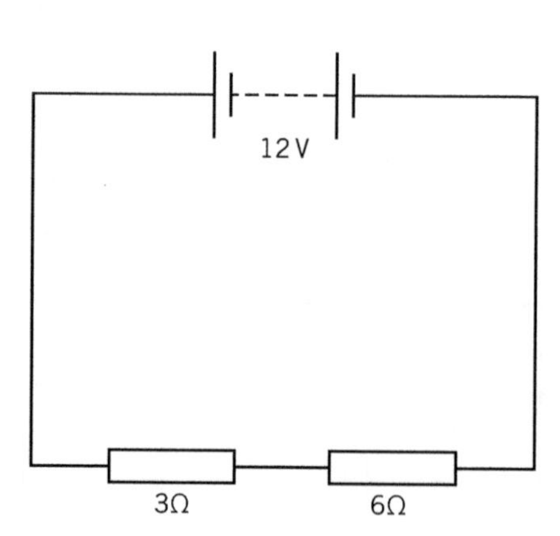

a Calculate the total resistance in the series circuit. [1 mark]

____

____

____

Total resistance = ________ Ω

b Explain why the resistance in the parallel circuit will be lower than the value calculated in part **a**. [3 marks]

Hint

This question is really just asking you to explain the difference between resistance in parallel and series circuits.

7 (H) A circuit was set up and run for 3 minutes, during this time the energy in the circuit was recorded as 45 J and the charge was 9 C. Calculate the resistance in the circuit. [4 marks]

Resistance = ______________ Ω

16 *I–V* characteristics

Investigate the effect of potential difference across a component on the current flowing through it.

Method

1. With the power supply switched off, set up the circuit shown in Figure 1.
2. You will adjust the variable resistor and measure the current and potential difference for your component. Do not allow the current to go above 1.0 A.
3. Starting with the variable resistor at its lowest resistance (so that the current is at its highest), measure the current and potential difference for your component.
4. Switch off the power supply.
5. Increase the resistance of the variable resistor in about six steps between the minimum and maximum resistances, and each time measure the current and potential difference for the component. Switch off the power supply between readings.
6. Reverse the polarity of the power pack by swapping the positive and negative connections and repeat steps 1-5.
7. Repeat the experiment for each component you are testing. When testing a diode, you should insert a fixed resistor in series with the diode, and swap the ammeter for a milliammeter.

Safety

- Components may get hot after being on for a while, so you should not touch them.
- Do not allow the current to go above 1.0 A, as this could cause overheating.
- Always switch off the power supply or disconnect the batteries before building or changing your circuit, and switch off the power supply between measurements.

Equipment

- power supply or battery pack
- components to test:
  - diode
  - filament lamp
  - resistor
- variable resistor
- ammeter
- voltmeter
- connecting leads

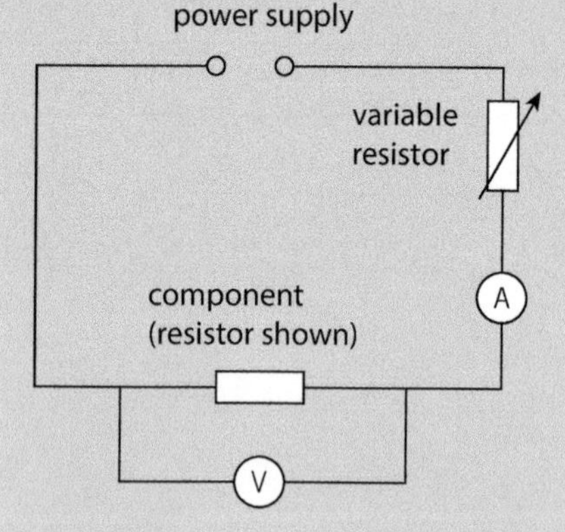

Figure 1

Remember

This experiment compares the way the current flowing through a component changes when the potential difference across that component is varied. The skills being tested are your ability to accurately measure current and potential difference, and whether you can understand and draw circuit diagrams. The results can be plotted on a graph of current (*I*) against potential difference (*V*). The particular shape of each *I–V* graph is described as the characteristic of the component.

Exam Tip

Filament bulbs, resistors, and diodes all have unique I–V characteristics. You should be able to recognise or sketch each component's I–V graph. You will also need to be able to explain their shape. You need to remember this equation for the resistance practical, but it is really important here as well:

$$\text{resistance }(\Omega) = \frac{\text{potential difference (V)}}{\text{current (A)}}$$

1 Draw one line from each circuit symbol to its name. [4 marks]

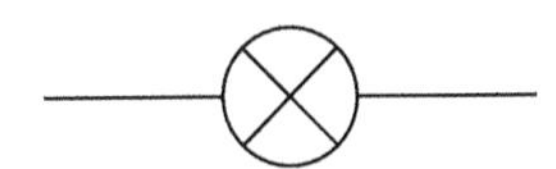

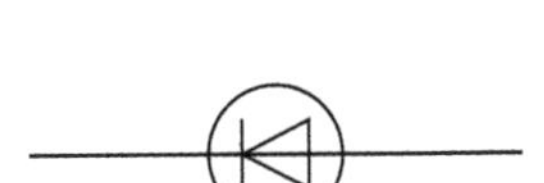

Battery
Cell
Diode
Filament lamp
Thermistor
Variable resistor

2 a i Sketch the current–potential difference graph for a filament lamp. [3 marks]

Exam Tip

When 'sketch' is the command word, you need to show the axis, the shape of the graph, and any coordinates where the line crosses the axes (if relevant). Sometimes the axes will be provided. You don't need a scale on the axes in a sketch graph.

ii Explain the shape of the current–potential difference graph for a filament lamp. [4 marks]

____

____

____

____

____

____

3 A student investigates the *I*–*V* characteristics of two unknown components. Their results are shown below.

Potential difference in V	Current in A Component 1	Current in A Component 2
−3.0	−0.14	0.00
−2.0	−0.12	0.00
−1.0	−0.05	0.00
0.0	0.00	0.00
1.0	0.05	0.01
2.0	0.11	0.18
3.0	0.15	0.40

a Plot both sets of data on the axes provided below. [4 marks]

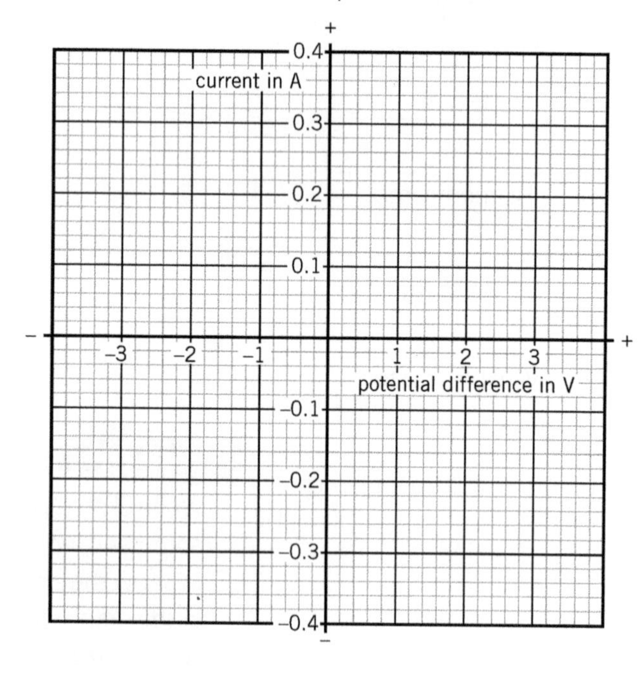

b Identify the components the student was testing. [2 marks]

- Component 1 = ____________________
- Component 2 = ____________________

c Compare and explain the shapes of the current–potential difference graphs for the two components. [6 marks]

____

____

____

____

____

____

____

____

____

____

d Use the graph to predict the current through component 1 when the potential difference across it is 4.0 V. [1 mark]

____

____

Current = ______ A

e Use the graph for component 1 to find its resistance. [3 marks]

____

____

Resistance of component 1 = ______ Ω

4 A student carried out two tests on a resistor with a fixed value. Their results are recorded below.

	Test 1	Test 2
current in A	14.8	9.2
potential difference in V	7.4	

Hint

You don't need to know the resistance to be able to answer this question – it's simply ratios.

a Use the values for test 1 to calculate the value of the resistor. [1 mark]

____

____

Resistance = __________ Ω

b Complete the table with the final measurement of potential difference. [2 marks]

____

____

____

5 Explain why *I–V* characteristics always go through the origin. [2 marks]

____

____

6 In this experiment, describe the purpose of reversing the battery. [2 marks]

____

____

7 Suggest an advantage of using a milliammeter instead of an ammeter. [1 mark]

____

____

17 Density

Investigate the density of regular- and irregular-shaped solids and liquids.

Method

A Density of a solid (in the shape of a cube or cuboid)

1. Measure and record the mass of the solid.
2. Measure and record the length, width, and height of the solid.

B Density of a solid (irregular-shaped sample)

3. Measure and record the mass of the sample.
4. Fill a measuring cylinder half-full with water. (There needs to be enough water so that when you put the solid into the water, the water will cover the solid but will not rise above the top of the measuring scale.)
5. With your eye level with the water's surface, measure the volume of water and record it.
6. Carefully place the solid material into the container.
7. Measure and record the new volume of the water.

C Density of a liquid

8. In your table for liquids, write the type of liquid whose density you are measuring.
9. Measure and record the mass of an empty measuring cylinder.
10. Pour the liquid into the measuring cylinder, making sure it does not go above the top of the cylinder's measuring scale.
11. Measure and record the mass of the cylinder with the liquid in it.
12. Measure and record the volume of liquid in the cylinder.

Equipment

- regular-shaped solid material
- irregular-shaped solid material
- liquid in a regular-shaped container
- balance
- ruler (30 cm is long enough)
- measuring cylinder

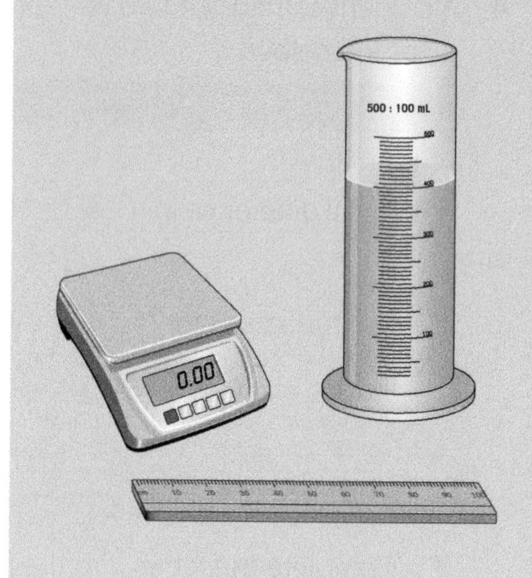

Safety

- Clean up any water spills straight away, reduce the risk of slipping.

Remember !

There are three separate experiments in this practical. You should be able to describe the methods for all three. This practical tests your ability to make accurate measurements of dimensions, mass, and volumes.

Exam Tip

You must know the equation linking density, mass, and volume.

Remember that the symbol for density is ρ. This is the lower case Greek letter rho, and is not a p (that is power or momentum). Don't get confused between the two symbols.

Converting between cubed units is not simple and requires practice. Remember that a 1 m^3 cube is not 100 cm^3. It's 100 cm on each side.

$1\,m^3 = 100\,cm \times 100\,cm \times 100\,cm = 1000000\,cm^3 = 10^6\,cm^3$

1 Draw one line from each object to the most appropriate set of equipment needed to safely find its density. [3 marks]

Object		Equipment
Pebble		Balance and a measuring cylinder
Textbook		Balance and a ruler
Battery		

2 When an object is placed into a displacement can, water pours out. Describe what the volume of water is equal to. [1 mark]

____

____

3 Suggest why it is important that an object is completely submerged before you measure the volume of water displaced. [1 mark]

____

____

4 A block of clay is found to have a volume of 50 cm^3 and a mass of 65 g.

a Write the equation that links density, mass, and volume [1 mark]

b Calculate the density of the block. [2 marks]

____

____

____

Density of clay block = __________ kg/m^3

Exam Tip

Always check to see if you need to convert any units before putting them in an equation.

5 **a** Calculate the volume of a $0.5\,m^3$ cube in cm^3. [2 marks]

____

____

Volume of block = ________ cm^3

Hint

The answer is not $50\,cm^3$. Remember that this is a three-dimensional shape!

b The block in part **a** has a density of $4.2\,g/cm^3$.

Calculate the mass of the block in kg. [3 marks]

____

____

Mass of block = ________ kg

6 Give two possible sources of error in this experiment. [2 marks]

____

____

7 100 ml of water has a mass of 100 g.

60 ml of honey has a mass of 87 g.

Identify which liquid is denser. [3 marks]

Exam Tip

If there is data in the question, always try to use it in your answer.

____

____

____

____

____

____

Answer = ________

8 A chess piece is carved from a regular cylinder of wood with a diameter of 4 cm and a height of 5 cm.

a Calculate the volume of the cylinder (use $\pi = 3.14$). [2 marks]

Hint

Volume of a cylinder = $\pi r^2 h$

____

____

____

Volume of cylinder = ________ cm^3

b Describe how you could use the apparatus pictured below to find out how much wood has been removed from the cylinder when carving the chess piece. [6 marks]

displacement can
water
measuring cylinder
object
overflow water

____

____

____

____

____

____

____

____

____

____

c A student weighs the cylinder before and after it was carved. Its mass was 44 g before carving and 33 g after it was carved.
Estimate the volume of the chess piece. [3 marks]

Hint
This question just needs you to use ratios to work out the answer. You will need your answer from part **a**.

____

____

____

____

____

Volume of finished chess piece = ________ cm^3

9 An irregular-shaped object has a recorded mass of 5.47 g.

The displays of three balances are shown below.

A	B	C
000.0 g	00.00 g	0.000 g

a Choose the balance which would be most appropriate to check the accuracy of the recorded mass. Tick **one** box. [1 mark]

A ☐

B ☐

C ☐

b Justify your answer to part **a**. [2 marks]

____

____

____

Hint

You may need to say why the other two balances are not as appropriate as your choice.

10 A student wants to identify an unknown liquid chemical in a bottle. Table 1 shows data the student recorded. Table 2 shows the densities of some common liquids.

Table 1

Mass of empty measuring cylinder	64.6 g
Mass of measuring cylinder with unknown liquid	142.8 g
Volume of unknown liquid	85 cm^3

Table 2

Liquid	Density g/cm^3
acetone	0.79
olive oil	0.92
petroleum	0.69
turpentine	0.87
water	1.00

Determine which of the liquids in Table 2 is most likely to be the unknown sample. [3 marks]

____

____

____

____

____

Unknown sample = ______________________________

11 A student heats some ice cubes in a beaker using a Bunsen burner. Describe the changes in the density of the water in terms of mass and volume. [3 marks]

____

____

____

____

____

Exam Tip

If you really understand the equations in physics. They can tell you a lot of information that can be used to answer longer written questions as well as maths questions.

12 The density of water is 1000 kg/m^3.

Object A has a density of 492 kg/m^3.

Object B has a density of 3673 kg/m^3.

Explain what will happen to objects A and B when they are placed in a beaker of water. [3 marks]

____

____

____

____

____

____

13 Calculate the mass of an object that has a density of 9300 kg/m^3, a depth of 20 cm, a height of 15 cm and a width of 42 cm. [3 marks]

____

____

____

____

____

____

Mass of object = __________ kg

18 Force and extension

Investigate the relationship between force and extension of a spring.

Method

1. Attach the spring to the clamp stand by hanging it off a clamp, and let the spring hang freely over the side of the bench.
2. Use the G-clamp to fix the clamp stand to the bench.
3. Use the other two clamps to hold the ruler vertically, near but not touching the spring. The marker pin should be attached to the hanger so that it lines up with the ruler. You will use this to measure the extension of the spring.
4. Record the point on the ruler that the marker pin points to with no mass attached. Record the force as 0 N.
5. Hang the 100 g mass holder (1.0 N force) from the spring and record the point on the ruler that the marker pin now points to.
6. Add a 1.0 N (100 g) weight and again record the point that the marker pin points to on the ruler.
7. Repeat step 6 until a total of 600 g (including the mass holder) is hanging from the spring.
8. Calculate the extension by taking the initial ruler measurement (step 4) away from the ruler measurement for each force applied to the spring.

Equipment

- eye protection
- spring
- set of 100 g masses
- 100 g mass holder
- 1 m ruler
- clamp stand with three clamps
- G-clamp
- marker pin

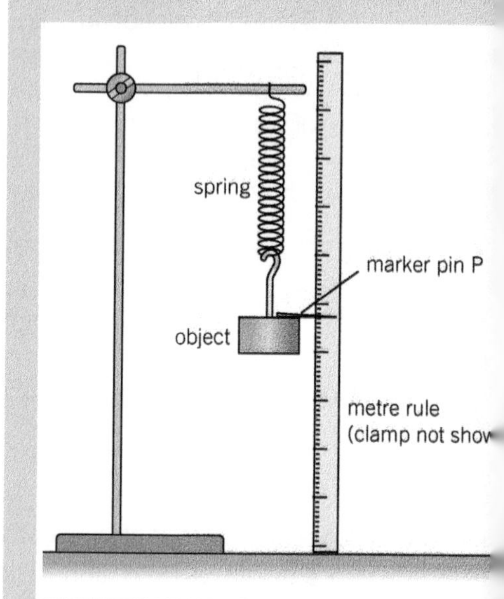

Safety

- Eye protection must be worn.
- The clamp stand should be securely fixed to the bench.
- Do not make the masses bounce up and down on the spring.
- Make sure masses do not fall onto the floor or onto people's feet.

Remember

This practical tests your ability to set up and use apparatus to accurately measure how much a spring extends by when you apply different forces by hanging masses from the spring. You should be able to label a diagram of the apparatus and describe the function of each piece of equipment.

You also need to be able to plot a force–extension graph from experimental results and you should know how to use this graph to calculate the spring constant.

The method described here uses a marker pin, but you may have carried out a method where you measured the length of the spring at each stage. You should be prepared to answer questions about both ways of carrying out the experiment.

Exam Tip

There are three equations related to this practical:

$F = ke$

Force (N) = spring constant (N/m) × extension (m) (need to be able to recall)

$E_e = \frac{1}{2}ke^2$

Elastic potential energy = ½ × spring constant (N/m) × extension^2 (m) (provided on data sheet)

$W = mg$

Weight (N) = mass (kg) × gravitational field strength (N/kg) (need to be able to recall)

Force, weight, and load are all terms used to mean the same thing in this practical, i.e., how much force is pulling on the spring.

1 Name the dependent variable in the experiment described in the method. [1 mark]

Dependent variable = ______________________________

2 A student hangs a 200 g mass from a spring. Calculate the weight acting on the spring. ($g = 9.8$ N/kg) [2 marks]

____

____

Mass = ______ N

Exam Tip

You need to be able to remember and apply the formula:

weight = mass × gravitational field strength

3 A student investigated the extension of a spring with increasing weight loaded on it. Their results are shown in the table below.

Force in N	Length of spring in cm	Extension of spring in cm
0	9	0
1	13	
2	15	
3	17	
4	19	
5	21	
6	23	

a Complete the table. [1 mark]

b
- Plot the data on the axes provided below.
- Draw a line of best fit. [3 marks]

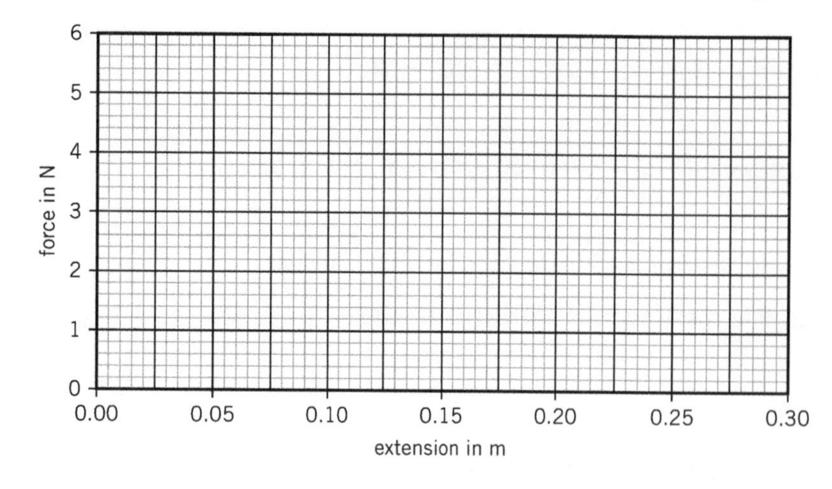

c Use the graph to name the type of relationship between weight and spring extension. [1 mark]

____

____

d Predict what might happen to the shape of the graph if more weight was added. [3 marks]

____

____

____

____

____

____

e Calculate the percentage change in length of a spring loaded with 6 N compared to an unstretched spring. [3 marks]

____

____

____

% change in length = __________ %

f Use the graph to calculate the spring constant. [3 marks]

____

____

____

____

Sprint constant = ______________ N/m

4 Suggest why it is important that safety glasses are worn in this practical. [1 mark]

____

____

5 Describe the function of the 'pointer' in this experiment. [1 mark]

____

____

6 Look at the diagram below.

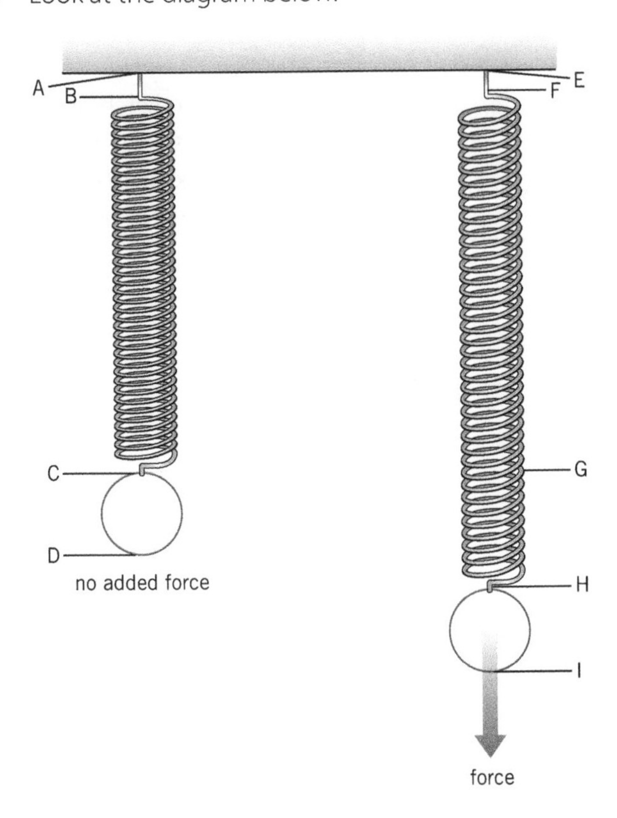

a Identify which points the initial length of the spring should be measured between.

Tick **one** box. [1 mark]

A and C ☐ B and C ☐

A and D  B and D 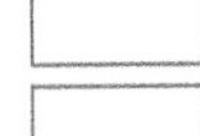

b Choose which points the length of the extended spring should be measured between. [1 mark]

Answer = ______ and ______

c Two students realise that they have measured all their lengths between the wrong two points (E and I).

- Student A says this will affect the accuracy of their results.
- Student B thinks that it will affect the precision of their results.

Explain which student is correct. [3 marks]

Exam Tip

You are being asked to explain and the question is worth three marks. This means you need to pick the correct student, give a reason why they are right, and possibly a reason why the other student is wrong.

7 A student carried out this experiment to compare the extension of two different springs.

One spring was very stiff and the other was easy to extend.

Describe the differences you would expect in the graphs force (*x*-axis) against extension (*y*-axis) for these two springs. [2 marks]

Exam Tip

You can use a sketch graph to aid you answer if you wish, but make sure you add helpful labels or annotations.

8 A trampoline manufacturer is investigating the spring constant of one of their trampoline springs by increasing the force pulling it and measuring the extended length.

a Explain why it is better to measure the length of the whole spring than just measuring the extension of the spring directly. [2 marks]

b Describe the changes in energy stores as weight is applied to the spring. [1 mark]

c The graph below shows the results of the manufacturer's experiment.

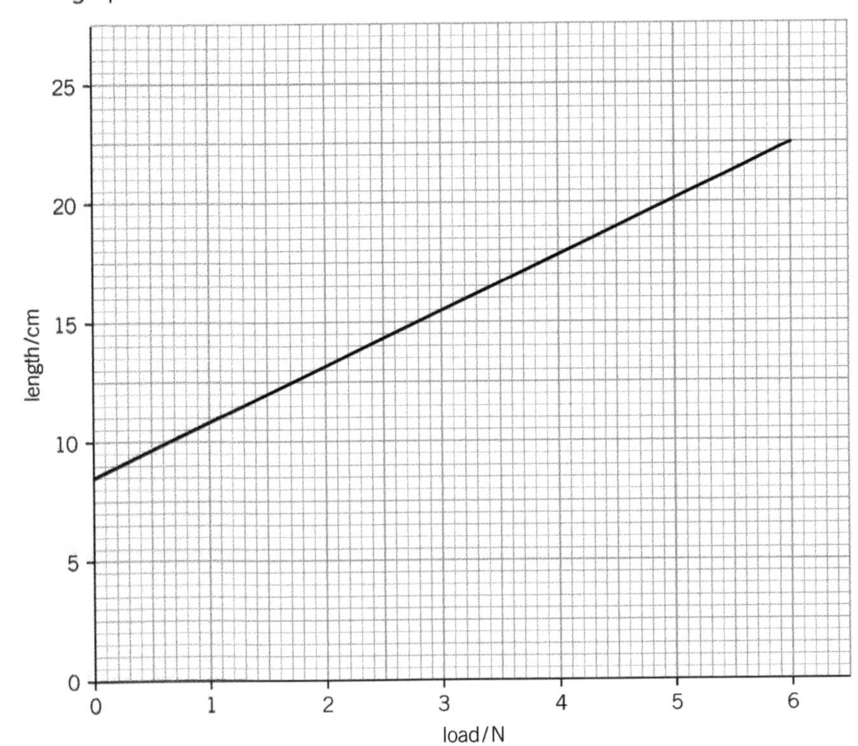

Describe the steps you would take to calculate the mass that would produce and extension of 5 cm. [3 marks]

Exam Tip

Don't worry about the fact it is a spring from a trampoline – it is still just a spring! Questions will often try to make you apply your knowledge to a situation or object you have not encountered before.

19 Acceleration

Investigate the relationship between force, mass, and acceleration.

Method

A The effect of force on the acceleration of a constant mass

1. Set up the apparatus as shown in Figure 1, with five chalk lines at 20 cm intervals. The distance from the start point to the edge of the bench should be 100 cm.
2. Check that the string is parallel to the bench.
3. Load the mass holder with 100 g and hold the trolley at the start point.
4. Set the stopwatch to lap mode.
5. Release the trolley at the same moment as starting the stopwatch.
6. Press the lap timer each time the trolley passes one of the chalk marks.
7. Record the times in a table.
8. Repeat steps 5 to 7 for masses of 80 g, 60 g, 40 g, and 20 g. To keep the mass of the system constant, you should stick the masses you remove from the holder on the trolley.

B The effect of an object's mass on its acceleration

9. Use the results of experiment A to choose a suitable weight to accelerate the trolley. Load the hanger with this mass.
10. Stick 100 g of masses on the trolley using the sticky tape.
11. Repeat steps 4-7 of experiment A.
12. Repeat steps 10 and 11 another five times increasing the mass by 100 g each time.

Equipment

- trolley
- ruler
- sticky tape
- chalk
- pulley with clamp
- mass holders
- 50 g masses
- 100 g masses
- string or strong thread

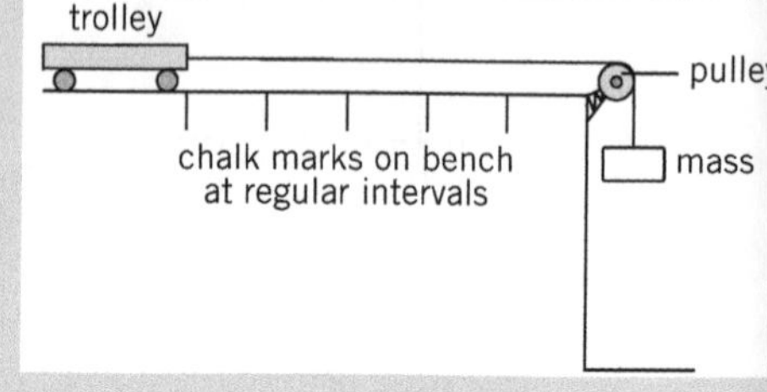

Figure 1

Safety

- The pulley must be safely attached to the bench.
- Masses should be securely attached to the trolley.
- Make sure no one is standing where the mass holder or masses could land on their feet.
- Make sure the trolley does not fall off the table.

Remember

You investigated whether the amount of force pulling an object changed its acceleration. You also investigated whether the mass of the trolley affected its acceleration when the force applied was kept constant.

Remember that this practical is testing your ability to observe the effect of a force by making and recording measurements of length, mass, and time. You should know how to use these measurements to calculate speed and acceleration.

Exam Tip

There are a lot of equations that you need to learn for this practical. Make sure you know which ones are given to you on the equation sheet in the exam, and which ones you need to learn. Even if they are given in the equation sheet, you should practise applying and rearranging them.

You could be expected to use a combination of two equations in one question.

1 Use words from the box below to complete Newton's second law. [3 marks]

acceleration	energy	forces	mass	speed	weight

An object's ______________ depends on the ______________ acting on the object and the ______________ of the object.

2 A student was testing how force affected the acceleration of a trolley with a mass of 800 g.

Exam Tip

Always work in kg.

Calculate the force acting on the trolley when the acceleration was 5 m/s^2. [2 marks]

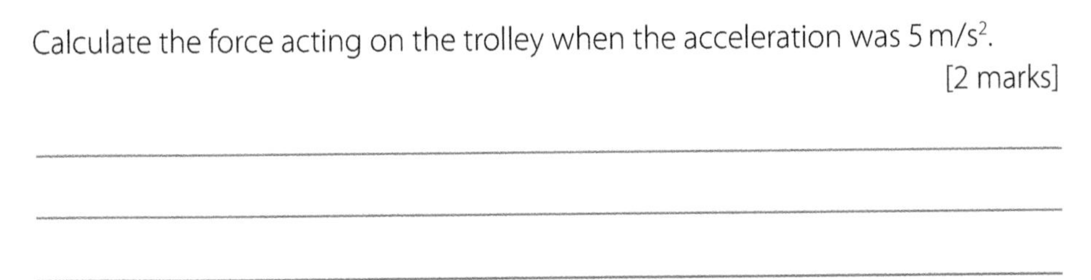

____

____

____

Force acting on trolley = ________ N

3 Two students carried out this experiment.

- Student A used the apparatus in Figure 1.
- Student B used the apparatus in the diagram below.

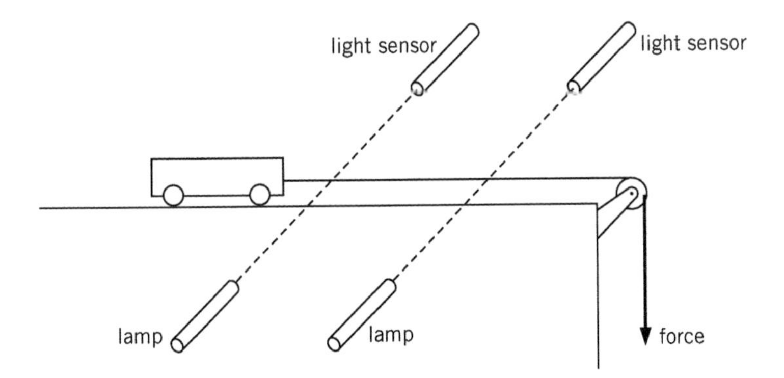

a Evaluate the two students' methods. [4 marks]

Exam Tip

An 'evaluate' question wants you to give evidence for and against. This could be based on information in the question, or on your own knowledge.

____

____

____

____

____

____

3 b Give a potential source of error in student A's method and suggest how they could record more accurate data without using light gates. [2 marks]

____

____

____

____

c What is the function of the pulley this experiment? [1 mark]

Tick **one** box.

Control direction of the trolley ☐

Prevent the trolley falling off ☐

Protect feet ☐

Reduce friction ☐

4 A student drew a graph of their data. The trolley used had a mass of 1.25 kg.

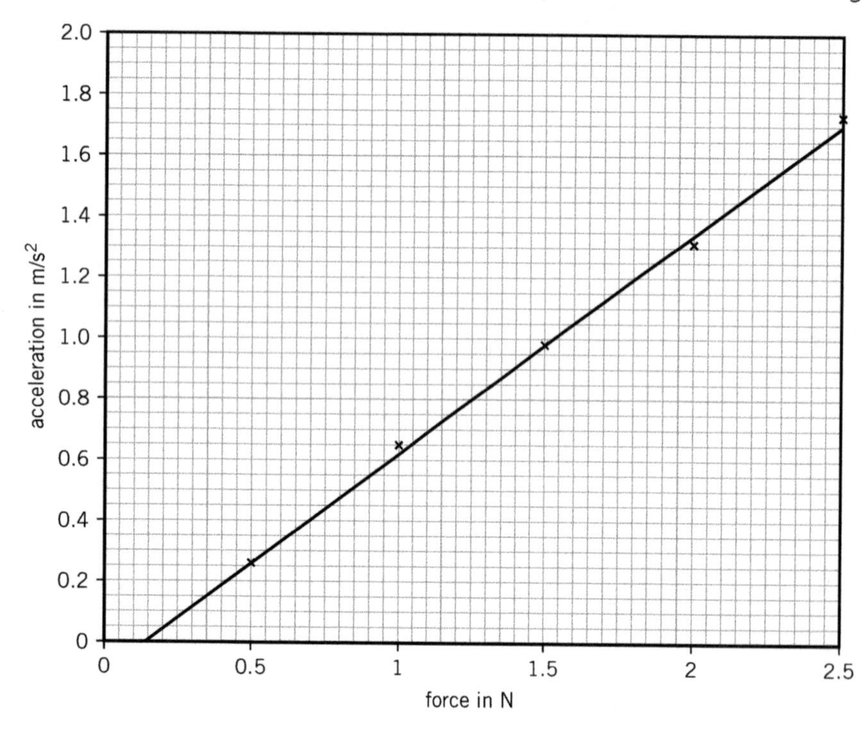

a Calculate the theoretical acceleration if the force is 1.5 N. [1 mark]

____

____

Acceleration = ________ m/s^2

b The student observed that the line of best fit doesn't go through the origin and the values they measured were always lower than the calculated theoretical acceleration.

Explain the student's observations. [3 marks]

____

____

____

____

____

5 Draw a free-body diagram to show the forces acting on the trolley in this practical. [4 marks]

Exam Tip

Remember this is science, not art. The car or trolley can be a simple as a dot with a label. Do not spend a lot of time drawing a car or a trolley.

Indicate the strength of the force by the relative size of the arrows and make sure you include labels.

6 There are a number of forces acting upon the object in this practical. Classify the following forces as either contact or non-contact forces. [5 marks]

friction gravity air resistance push pull

Contact force	Non-contact force

7 Describe the difference between speed and velocity. [2 marks]

____

____

____

____

8 A group of students tested how acceleration was affected by changing the force acting on a trolley of constant mass.

Force in N	Acceleration in m/s^2
2	0.98
4	1.92
6	2.93
8	3.92
10	4.70

a Plot the data on the axes provided. [3 marks]

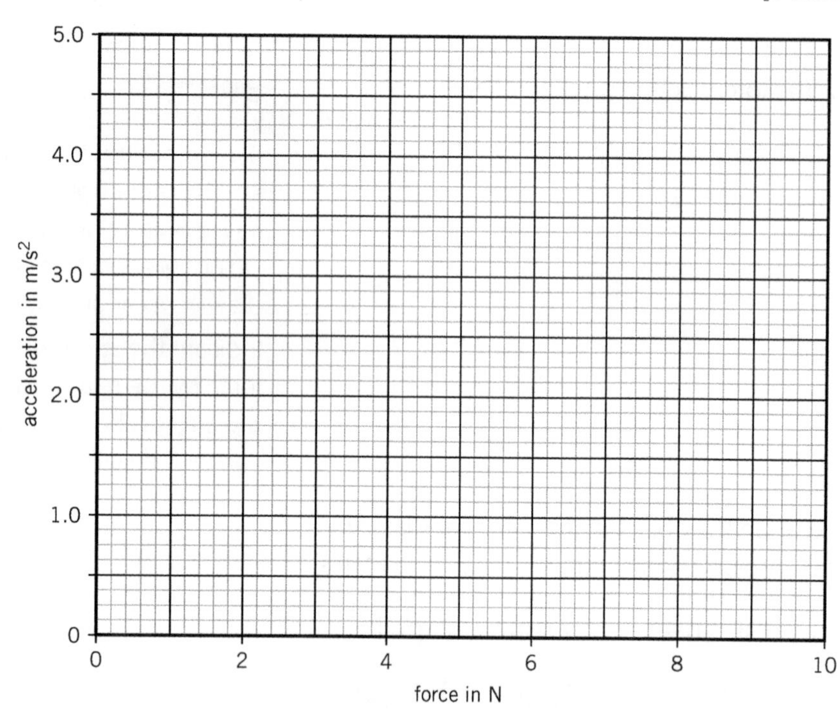

b Describe the type of relationship seen in the graph. [1 mark]

____

____

Another student changed the mass but kept a constant force acting on the car.

Mass of car in kg	Acceleration in m/s^2
2	4.92
4	2.46
6	1.64
8	1.23
10	0.99

c Use the data to estimate a value for the resultant force accelerating the car. [2 marks]

____

____

____

Resultant force = ____________ N

9 An accelerating trolley of mass 1.5 kg passed one light gate with a velocity of 4.5 m/s.

The trolley passed a second light gate 3 seconds later with a velocity of 5.1 m/s. Calculate the resultant force acting on the trolley. [4 marks]

Resultant force = ________ N

10 A student uses a single light gate to record the velocity of a trolley 30.0 cm from its starting position. The trolley begins at rest and the velocity is recorded as 3.87 m/s. The equation for uniform acceleration is:

$$v^2 - u^2 = 2as.$$

Exam Tip

This equation is given in the Physics equation sheet in the exam, but you need to know how to apply it.

Remember that *s* means distance – not speed!

a Rearrange the equation to make *a* the subject of the formula. [1 mark]

b Calculate the acceleration of the trolley. [2 marks]

Acceleration of trolley = ________ m/s^2

c The trolley has a mass of 200 g.

Estimate the mass of the hanger that is pulling on the trolley.

(g = 9.8 N/kg) [5 marks]

Mass of hanger = ________ g

20 Waves

Investigate the behaviour of waves in solids and liquids.

Method

A Wave behaviour in a solid

1. Assemble the apparatus as shown in Figure 1.
2. Measure the length of string between the vibration generator and the pulley.
3. Turn on the signal generator.
4. Increase the frequency of the vibration from zero until you can see one complete wave on the string.
5. Record the frequency of the signal generator.
6. Continue to increase the frequency until you see a second wave pattern, with two complete waves.
7. Record the frequency on the signal generator.
8. Continue this process to find the frequencies which form three and four complete waves on the string.

B Wave behaviour in a liquid

1. Set up the ripple tank as shown in Figure 2.
2. Fill the tray with water about 1 cm deep. The dipper should just touch the surface of the water.
3. Place the lamp so that it shines into the tray.
4. Turn on the power supply and lower the speed until separate waves are clearly visible on the screen below.
5. Measure the distance between the two furthest apart visible waves on the screen. Divide this distance by the number of waves visible to calculate the wavelength.
6. Count the number of waves passing a fixed point over a period of 10 seconds. Divide this number by ten to find the frequency.

Equipment

- signal generator
- oscillator
- masses on a mass holder
- length of string
- tape measure
- ripple-tank
- low-voltage power supply
- lamp

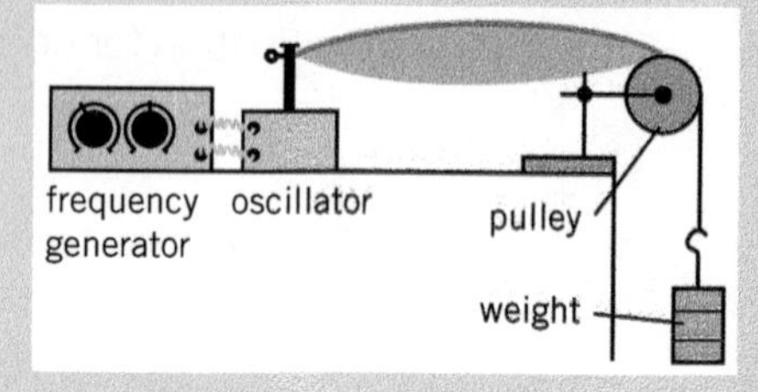

Figure 1

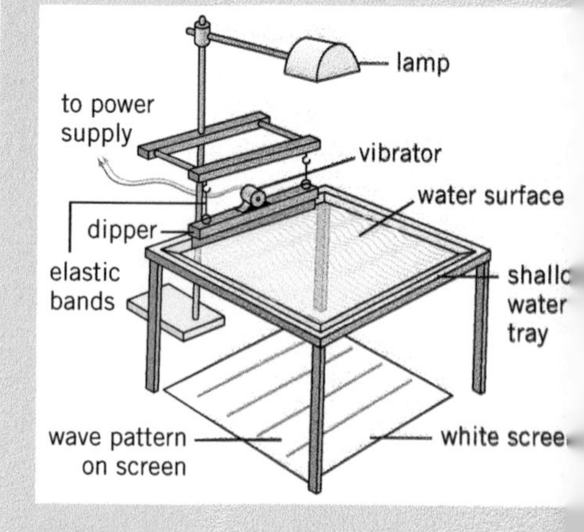

Figure 2

Safety

- Do not touch the vibration generator when it is in motion.
- Be careful not to drop masses on the floor.
- Clean up any water spill immediately.
- If using a lamp, keep the lamp as far away from the water as possible.
- Do not touch the lamp or electrical supply with wet hands.

Remember !

This practical tests your ability to use information from very different types of equipment to investigate the properties of waves in solids and in liquids. The physics and the theories are the same in both experiments – they both allow speed, frequency, and wavelength to be measured.

Exam Tip

There are a few equations you need for this practical. You need to learn and know how to apply the following equations.:

$v = f\lambda$

wave speed (m/s) = frequency of wave (Hz) × wavelength (m)

$s = vt$.

distance (m) = speed (m/s) × time (s).

$$\text{frequency} = \frac{\text{number of waves passing a fixed point}}{\text{Time}}$$

1 Figure 1 shows the apparatus for investigating the behaviour of a wave on a string.

a Sketch the wave produced on the string and label it with the amplitude and the wavelength. [3 marks]

____

____

____

____

____

____

b Describe the direction of movement of particles and transfer of energy in the wave on the string. [2 marks]

____

____

____

2 A student pushes a slinky back and forth to investigate longitudinal waves. Use the words provided to label the diagram of the student's experiment. [3 marks]

compression	rarefaction	wavelength

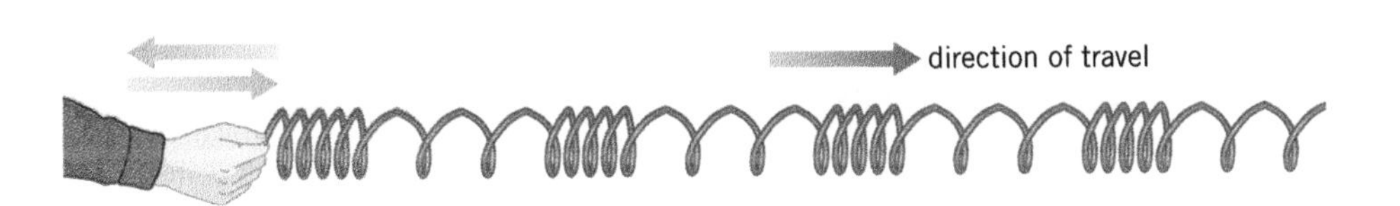

3 When using a ripple tank, the wave may travel very quickly making it difficult to measure the wavelength.

Setting up a strobe light can make the wave appear stationary and make them easier to count and measure.

a Suggest an alternative way to measure the wavelength accurately that does not involve a strobe light. [3 marks]

____

____

____

____

b Explain the function of the white screen and the lamp. [2 marks]

____

____

____

c Using the ripple tank, a student counted 22 waves passing a fixed point in 5.0 seconds.

Calculate the frequency of the waves in the ripple tank. [2 marks]

Frequency of waves = ________ Hz

d The student measured the wavelength of the waves in part **b** to be 0.03 m.

Calculate the wave speed and select the correct units from the box below.

[4 marks]

Hz	m	m/s	m/s^2	s

____

____

____

Wave speed = ____________ Units = ____________

4 The diagram below shows what a student observed when investigating a wave on a string.

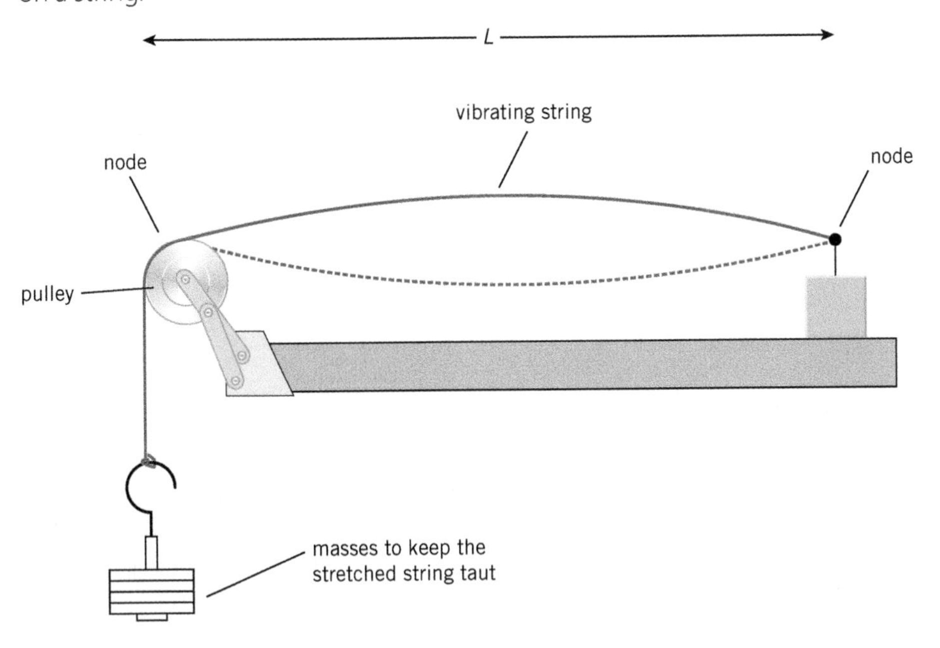

The student made the following claim.

'Wavelength can't be measured as there isn't a full wave on the string.'

Explain whether the student is correct. [3 marks]

____

____

____

____

5 Two students carried out an experiment measuring waves in a ripple tank.

- Student A measured the wavelength of a single wave.
- Student B measured the wavelength of ten waves and then divided the value by 10.

Explain which student's data will be the most accurate and why. [3 marks]

____

____

____

____

6 Give an advantage of using a wave generator or a motor to produce waves in a ripple tank, instead of producing them by hand. [1 mark]

____

7 A student carried out a version of the ripple tank experiment to investigate waves radiating out in a circle. A diagram of their experiment is shown below.

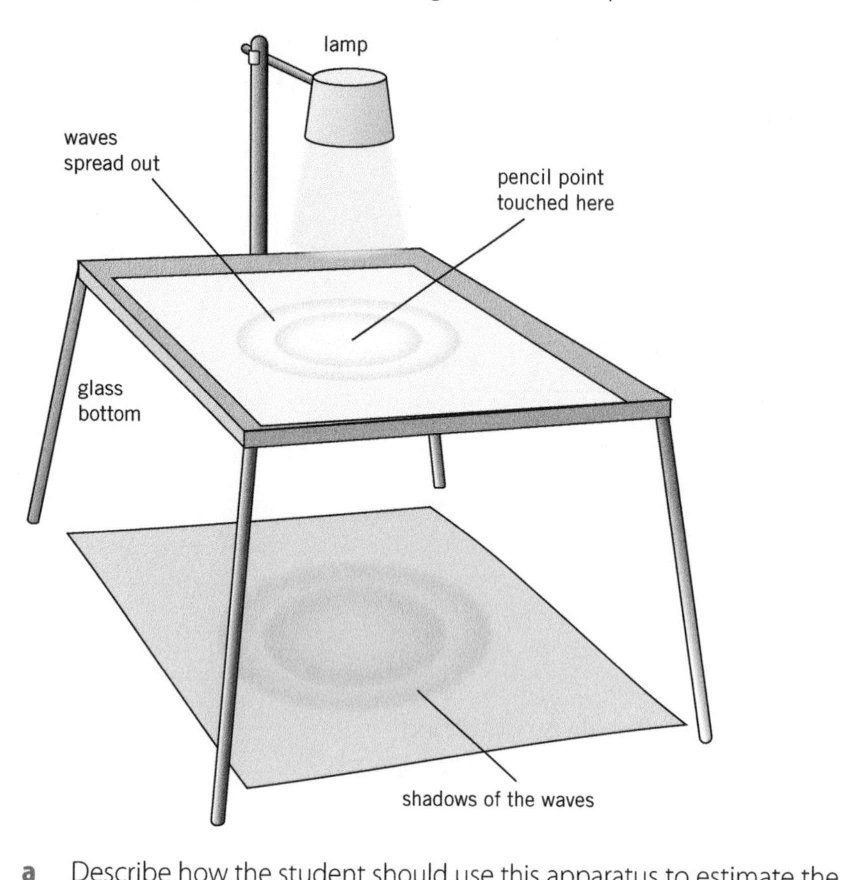

Exam Tip

This is a slightly different experiment because the waves are forming circles rather than straight lines. Don't be put off by the slight difference – the theory is exactly the same!

a Describe how the student should use this apparatus to estimate the wave speed as accurately as possible. [6 marks]

b The student counted 17 waves passing a fixed point in 11 seconds.

Calculate the frequency of the waves. [3 marks]

___

___

___

Frequency = ________ Hz

Exam Tip

There is often a mark for giving your answer to the correct number of significant figures. Remember that you should never give more significant figures than the original numbers in the question.

8 A scientist is investigating the properties of an electromagnetic wave in a vacuum. Their detector measures the frequency, f as 2.0×10^{10} kHz. The speed of electromagnetic waves in a vacuum is 3.0×10^{8} m/s.

Calculate the wavelength of the electromagnetic wave. [3 marks]

___

___

___

___

___

Wavelength = ______________ m

Hint

Pay close attention to the units given to you in the question. Do you need to convert anything?

21 Radiation and absorption

Investigate the amount of infrared radiated from different types of surfaces.

Method

A Emission of infrared radiation by silver and black cans.

1. Place the two drinks cans on a heatproof mat. Beakers with aluminium foil lids can be used as an alternative.
2. Carefully pour 150 ml of hot water into each of the two cans using a measuring cylinder.
3. If you are using beakers, place the lids onto the beakers.
4. Put a thermometer into the water in each container.
5. Start the stopwatch, and record the temperature of the water in the two containers every 30 s for 10 minutes.

B Emission of infrared radiation from a Leslie cube.

1. Place the Leslie cube on a heatproof mat and fill it with hot water.
2. Use an infrared thermometer/detector to measure the amount of infrared radiating from one surface of the Leslie cube.
3. Carefully rotate the Leslie cube to measure the amount of infrared radiating from each of the surfaces.
4. Record your results in a suitable table.

Make sure the infrared detector/thermometer is always the same distance from the surface being tested.

Equipment

- two drink cans (one painted silver and one painted matt black)
- two thermometers
- measuring cylinder
- stopwatch
- hot water
- heatproof mat
- heatproof gloves
- Leslie cube
- infrared thermometer or detector

Safety

- Wear eye protection in case of hot water splashes.
- Take care with hot water. Use heatproof gloves if you need to move hot water from one place to another and warn anyone that is nearby before you move.
- Clean up any water spills straightaway.
- Do not touch the hot water containers, as they will get very hot. Leave them to cool before clearing them away after the practical.

Remember

This practical is all about investigating how the type of surface affects the amount of infrared radiation that is emitted. There are lot of different methods for this practical. Some methods involve fancy equipment and others use very basic equipment. The science behind the experiment is the same no matter how you carried out the practical.

Exam Tip 

Applying your knowledge to unfamiliar situations is an important skill for you to practice as exam questions will usually try to present information in a way you haven't seen before. Try to work out how what you see relates to the experiment that you did.

1 a Complete the order of waves in the electromagnetic spectrum using the terms in the box below. [2 marks]

x-rays ultraviolet infrared microwaves

radio waves
visible light
gamma rays

b On the electromagnetic spectrum you completed in part **a**, circle the type of radiation that is emitted by hot objects. [1 mark]

c State the main source of this radiation on Earth. [1 mark]

____

d List two properties all electromagnetic waves have in common. [2 marks]

____

____

2 A student investigates the emission of radiation from a silver can and a matt black can as described in the method section.

a Describe three variables that should be controlled in this experiment. [3 marks]

____

____

____

____

The student's results are shown below.

Time in seconds	Temperature of water in °C	
	Silver can	Matt-black can
0	98	98
60	90	85
120	83	74
180	75	64
240	68	55
300	62	49

b Use the results to identify which can is the better emitter of infrared radiation and give a reason for your choice. [2 marks]

____

____

____

c Calculate the rate of temperature change for each can in °C/s. [4 marks]

____

____

____

____

____

____

Rate of temperature change for silver can = ________ °C/s

Rate of temperature change for matt-black can = ________ °C/s

d Calculate the % change in the temperature of the water in the silver can between 0 seconds and 300 seconds. [2 marks]

____

____

____

% change in the temperature of the water in silver can = ________ %

e The student also has an infrared lamp available.

Describe how the student would use this equipment to investigate which can is the best absorber of infrared radiation. [6 marks]

____

____

____

____

____

____

____

____

____

____

3 A student wants to investigate how the temperature of water in a can affects the amount of infrared radiated.

a Circle the correct options to complete the student's hypothesis below. [4 marks]

'I predict that the amount of **infrared/ultraviolet** radiation **emitted/absorbed** by the can will be **directly/inversely** proportional to the temperature of the water in the can.

This is because the **higher/lower** the temperature of an object, the more infrared radiation it emits.'

b Sketch a graph on the axes below to show the results that the hypothesis predicts. [2 marks]

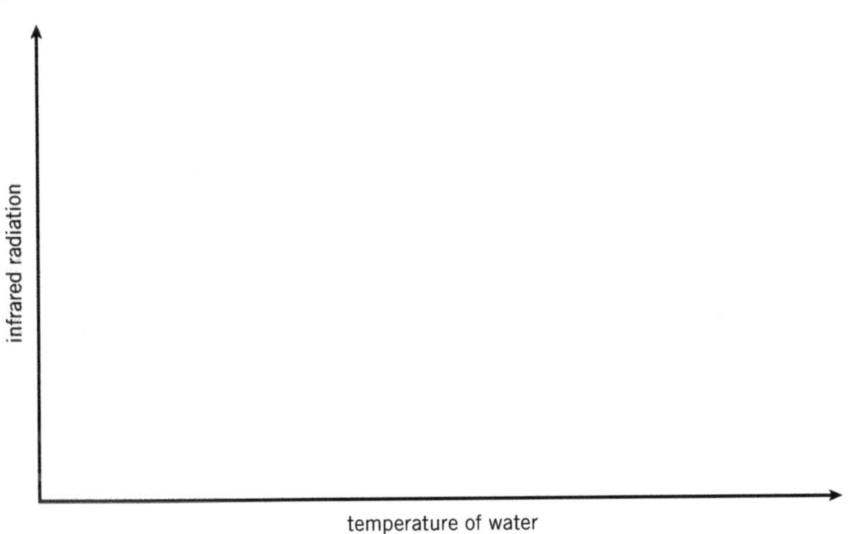

4 A student plans an experiment to measure the amount of infrared radiation emitted from different surfaces using the apparatus shown below.

[1 mark]

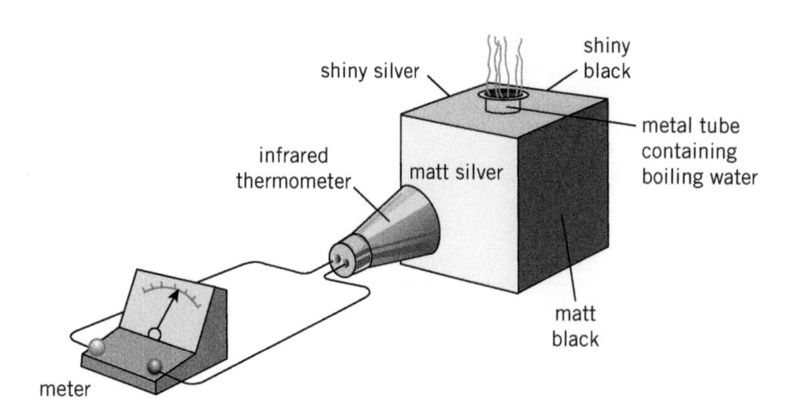

a Name the independent variable in this experiment. [1 mark]

____

____

Exam Tip

Time is precious in an exam so do NOT start you answer, 'in this experiment the independent variable is….'

Repeating the question won't gain you any marks – just give the answer.

A student measured the infrared radiation from the four different surfaces using the infrared detector.

Surface	Infrared radiation in units
shiny silver	17
matt silver	35
shiny black	39
matt black	47

Exam Tip

Some tables and graphs have variables that say 'in units' or 'in arbitrary units'. If you see this in an exam don't panic – they are just used to show a general trend or values when the actual units are not needed.

b Plot this data on the graph paper provided below. [4 marks]

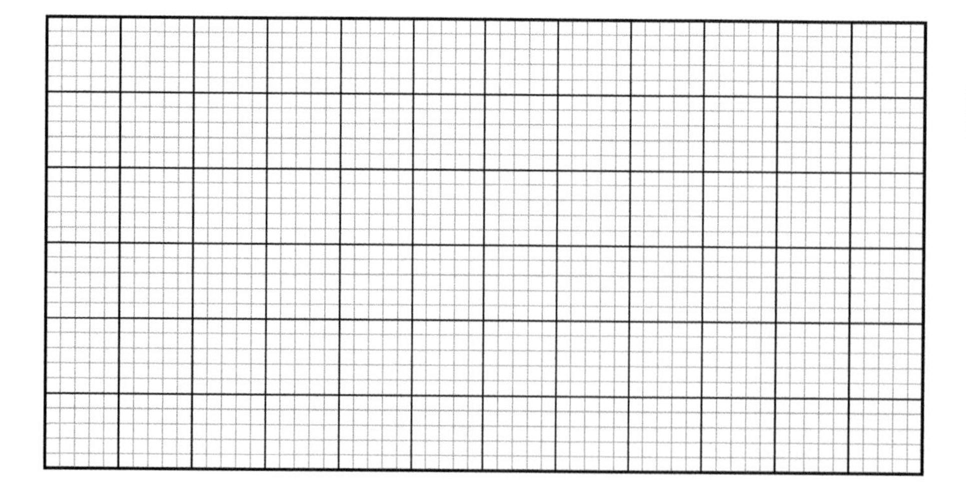

Exam Tip

Think carefully about the type of graph you draw when one of the variables is discrete data.

c The student's hypothesis was:

'The colour of the surface will affect emission of infrared radiation more than the roughness of the surface.'

Write a conclusion based on the student's results and their hypothesis. [4 marks]

Exam Tip

If you are asked to write a conclusion, always quote data to back up what you are saying.

d Predict which surface would be the best absorber and which surface would be the poorest absorber of infrared radiation and explain why in each case. [4 marks]

Hint

You could draw an annotated diagram to help answer this question.

Physics equations

You should be able to remember and apply the following equations, using SI units, for your assessments.

Word equation	Symbol equation
weight = mass × gravitational field strength	$W = mg$
force applied to a spring = spring constant × extension	$F = ke$
acceleration = $\frac{\text{change in velocity}}{\text{time taken}}$	$a = \frac{\Delta v}{t}$
H momentum = mass × velocity	$p = mv$
gravitational potential energy = mass × gravitational field strength × height	$E_p = mgh$
power = $\frac{\text{work done}}{\text{time}}$	$P = \frac{W}{t}$
efficiency = $\frac{\text{useful power output}}{\text{total power input}}$	
charge flow = current × time	$Q = It$
power = potential difference × current	$P = VI$
energy transferred = power × time	$E = Pt$
density = $\frac{\text{mass}}{\text{volume}}$	$\rho = \frac{m}{V}$
work done = force × distance (along the line of action of the force)	$W = Fs$
distance travelled = speed × time	$s = vt$
resultant force = mass × acceleration	$F = ma$
kinetic energy = 0.5 × mass × $(\text{speed})^2$	$E_k = \frac{1}{2}mv^2$
power = $\frac{\text{energy transferred}}{\text{time}}$	$P = \frac{E}{t}$
efficiency = $\frac{\text{useful output energy transfer}}{\text{total input energy transfer}}$	
wave speed = frequency × wavelength	$v = f\lambda$
potential difference = current × resistance	$V = IR$
power = current^2 × resistance	$P = I^2R$
energy transferred = charge flow × potential difference	$E = QV$

You should be able to select and apply the following equations from the Physics equation sheet.

Word equation	Symbol equation
(final velocity)^2 – (initial velocity)^2 = 2 × acceleration × distance	$v^2 - u^2 = 2as$
elastic potential energy = 0.5 × spring constant × extension^2	$E_e = \frac{1}{2}ke^2$
period = $\frac{1}{\text{frequency}}$	
H force on a conductor (at right angles to a magnetic field) carrying a current = magnetic flux density × current × length	$F = BIl$
change in thermal energy = mass × specific heat capacity × temperature change	$\Delta E = mc\Delta\theta$
thermal energy for a change of state = mass × specific latent heat	$E = mL$

key

relative atomic mass **atomic symbol** name atomic (proton) number

1	2											3	4	5	6	7	0
							1 **H** hydrogen 1										4 **He** helium 2
7 **Li** lithium 3	9 **Be** beryllium 4											11 **B** boron 5	12 **C** carbon 6	14 **N** nitrogen 7	16 **O** oxygen 8	19 **F** fluorine 9	20 **Ne** neon 10
23 **Na** sodium 11	24 **Mg** magnesium 12											27 **Al** aluminium 13	28 **Si** silicon 14	31 **P** phosphorus 15	32 **S** sulfur 16	35.5 **Cl** chlorine 17	40 **Ar** argon 18
39 **K** potassium 19	40 **Ca** calcium 20	45 **Sc** scandium 21	48 **Ti** titanium 22	51 **V** vanadium 23	52 **Cr** chromium 24	55 **Mn** manganese 25	56 **Fe** iron 26	59 **Co** cobalt 27	59 **Ni** nickel 28	63.5 **Cu** copper 29	65 **Zn** zinc 30	70 **Ga** gallium 31	73 **Ge** germanium 32	75 **As** arsenic 33	79 **Se** selenium 34	80 **Br** bromine 35	84 **Kr** krypton 36
85 **Rb** rubidium 37	88 **Sr** strontium 38	89 **Y** yttrium 39	91 **Zr** zirconium 40	93 **Nb** niobium 41	96 **Mo** molybdenum 42	[98] **Tc** technetium 43	101 **Ru** ruthenium 44	103 **Rh** rhodium 45	106 **Pd** palladium 46	108 **Ag** silver 47	112 **Cd** cadmium 48	115 **In** indium 49	119 **Sn** tin 50	122 **Sb** antimony 51	128 **Te** tellurium 52	127 **I** iodine 53	131 **Xe** xenon 54
133 **Cs** caesium 55	137 **Ba** barium 56	139 **La*** lanthanum 57	178 **Hf** hafnium 72	181 **Ta** tantalum 73	184 **W** tungsten 74	186 **Re** rhenium 75	190 **Os** osmium 76	192 **Ir** iridium 77	195 **Pt** platinum 78	197 **Au** gold 79	201 **Hg** mercury 80	204 **Tl** thallium 81	207 **Pb** lead 82	209 **Bi** bismuth 83	[209] **Po** polonium 84	[210] **At** astatine 85	[222] **Rn** radon 86
[223] **Fr** francium 87	[226] **Ra** radium 88	[227] **Ac*** actinium 89	[261] **Rf** rutherfordium 104	[262] **Db** dubnium 105	[266] **Sg** seaborgium 106	[264] **Bh** bohrium 107	[277] **Hs** hassium 108	[268] **Mt** meitnerium 109	[271] **Ds** darmstadtium 110	[272] **Rg** roentgenium 111	[285] **Cn** copernicium 112	[286] **Nh** nihonium 113	[289] **Fl** flerovium 114	[289] **Mc** moscovium 115	[293] **Lv** livermorium 116	[294] **Ts** tennessine 117	[294] **Og** oganesson 118

*The lanthanides (atomic numbers 58–71) and the actinides (atomic numbers 90–103) have been omitted.
Relative atomic masses for **Cu** and **Cl** have not been rounded to the nearest whole number.

Notes

Notes

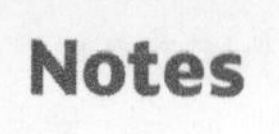